Roger Adams
Scientist and Statesman

D. Stanley Tarbell
Ann Tracy Tarbell

AMERICAN CHEMICAL SOCIETY
WASHINGTON, D. C. 1981

Library of Congress Cataloging in Publication Data

Tarbell, D. Stanley, 1913–
Roger Adams: scientist and statesman.

Includes bibliographies.

1. Adams, Roger, 1889–1971. 2. Chemists—United States—Biography.
I. Tarbell, Ann Tracy, 1916– . II. Title.

QD22.A317T37 540'.92'4 [B] 81–17625
ISBN 0–8412–0598–1 AACR2
ISBN 0–8412–0711–9 (paperback) 1981

PRINTED IN THE UNITED STATES OF AMERICA

Second printing 1983

Contents

Preface

During studies on the development of American organic chemistry, we talked with E. H. Volwiler about the growth of research at Abbott Laboratories. He mentioned that no one had written a comprehensive biographical sketch of Roger Adams, and that it would be a useful project to study his career in some detail. We were attracted by the idea, which we had considered tentatively before, because of our personal experience with Adams and because it would be complementary to our other studies. Lucile Adams Brink not only warmly welcomed Volwiler's suggestion of a biography of her father but also furnished us with family papers, photographs, and memorabilia, in addition to much oral and written information about her father and mother. Volwiler and Lucile Brink read and commented on our manuscript as it progressed, and we are grateful for their assistance and encouragement. We are indebted to R. M. Joyce for his critical reading and helpful suggestions and for his assiduous collecting of pertinent material.

The bulk of Adams's extant papers are excellently arranged and classified in the University of Illinois Archives at Urbana, and our debt to Archivist Maynard E. Brichford and to his staff for their help during numerous visits to Urbana is great. Jean R. St. Clair, archivist of the National Academy of Sciences, found key Adams documents from the academy files for us. At the National Archives Charles Dewing aided with documents throwing light on Adams's work in Washington from 1941 to 1945. The archives of Radcliffe College and Harvard University furnished information about the careers of Adams and his sisters at these institutions. The Conant Public Library of Winchester, N.H., supplied background material about the New Hampshire of Adams's progenitors. The university libraries of Illinois and Vanderbilt have been invaluable resources.

J. Merton England of the National Science Foundation sent us correspondence showing Adams's work with the National Science Board, and the American Chemical Society provided us with useful documents.

C. S. Marvel and H. E. Carter gave us much information during personal interviews and examined the complete manuscript. Many other people aided us by encouragement, by commenting on the manuscript, by furnishing correspondence, and by personal recollections of Adams. These include Richard T. Arnold, Sidney H. Babcock, John C. Bailar, Jr., Emma P. Baskerville, Karl M. Beck, A. Harold Blatt, Theodore L. Cairns, Ralph Connor, E. J. Corey, David Y. Curtin, Oscar H. Dodson, the late Robert C. Elderfield, John Edwards, C. Harold Fisher, the late Reynold C. Fuson, Henry Gilman, Cyrus O. Guss, Herbert S. Gutowsky, William E. Hanford, Karl Heumann, Nelson J. Leonard, William H. Lycan, J. C. Martin, Edward J. Matson, Koji Nakanishi, Wayland E. Noland, Mrs. William Palmer, Norman Rabjohn, William C. Rose, John D. Roberts, Edwin M. Shand, Jack Simpson, Harold R. Snyder, Ruth and William E. Tracy, the late Teresa Norton Turner, Lillian Volwiler, Beulah Westerman, Arthur W. Weston, Irene and William M. Wheeler, John A. Wheeler, Elsie Wilson, and Harold E. Zaugg. The responsibility for opinions expressed and for mistakes is, of course, ours.

Expenses for travel, for making copies of important sources, and for some typing were defrayed by research grants from the Centennial Fund of Vanderbilt University and the Petroleum Research Fund and by a grant from the National Science Foundation. Most of the typing of the many drafts was done by the secretarial staff of the Vanderbilt Department of Chemistry. We are indebted to all the people and agencies mentioned for their support and cooperation. We are also grateful to colleagues, both here and elsewhere, and to our families for their continuing interest and encouragement in this project.

That so many of Roger Adams's friends have been anxious to share information and memories of him to make the present work possible is an indication of the respect and affection he inspired.

D. Stanley Tarbell
Ann Tracy Tarbell

Department of Chemistry
Vanderbilt University

Introduction

Over September 3–4, 1954, the University of Illinois at Urbana–Champaign was the scene of a symposium, attended by over 500 people, of special significance for American and international science. Roger Adams, the leading organic chemist of his generation, an Illinois faculty member since 1916 and head of the chemistry department since 1926, was retiring as head at the age of sixty-five. He was to serve as research professor for three more years and to continue his active career as scientific statesman until his death in 1971. His friends planned this occasion to try to show him what he meant to them and to recognize his services, on a larger stage, to American chemistry and to international science. By 1954 Adams, "the Chief," as he was known to generations of students, had trained 172 Ph.D.'s and about 45 postdoctoral fellows and had published some 375 scientific papers. However, his students, colleagues, and friends had not flocked to Urbana from distant places merely to honor an outstanding organic chemist. They were drawn by a combination of respect and personal affection for a man who was unique in his time and whose stature has become clearer in the time since 1954.

Adams was generally believed to be related to the presidential Adams family of Massachusetts, although few people knew exactly how, and he never mentioned the matter. Probably everyone knew that he was the founder and guiding spirit of *Organic Syntheses* and *Organic Reactions*, although not many knew the story of these valuable publications. All knew that he had been a research consultant to several leading industrial research laboratories, but not many were aware of his unique role in the development of industrial research and technology in this country. His collaborators knew him as a person of great charm and force who was able to make interesting the drudgery encountered in research and to make positive results exciting. They knew that with his colleagues he had built up an outstanding graduate chemistry department, but not all realized the importance of his contribution to graduate education in science.

Most of those present knew that Adams had served his country as a scientist in two world wars, but few realized the significance of his scientific missions to

Germany and Japan in 1945–48. All knew of some of his public services to the American Chemical Society and the American Association for the Advancement of Science, but few understood the scope and value of his work for these and other scientific groups. In 1954 Adams was just starting to plan a new program for the Alfred P. Sloan Foundation; none could know that later recommendations for the support of research programs of promising younger men in the physical sciences would make the Sloan Foundation a valuable force in American science.

All of his friends at the symposium recognized the Chief's interest in them as individuals, but not all were aware of the full extent of this interest. All knew that Adams was at home with people of all walks of life, anxious to learn from them what they could impart. Like Kipling's hero, he walked with kings yet kept the common touch. He was equally interested in talking with the host of a tiny inn far back in the Austrian Tyrol, with a Dutch minister of education, with General Douglas MacArthur, and with the emperor of Japan.

However incomplete their knowledge of Adams's total accomplishments may have been, those gathered at Urbana in September 1954 came to honor a friend and a unique personality, with whom collaboration had been a unique privilege.

Adams's own Ph.D.'s showed their respect in the gift of a huge commemorative volume, containing a letter and a picture from each student.[1] The letters paid tribute to Adams in terms limited only by the powers of expression and personalities of the writers. The following are among the more striking:[2]

> To a teacher who has given his students, while they were at Illinois and in their later careers—
>
> —an everlasting interest in chemistry and a sense of the still untouched frontiers,
>
> —an example of how to work like hell,
>
> —a liberal education in the ways of the world,
>
> —and an appreciation of how to lose at poker and enjoy it.

> Knowing you has been one of the best things that ever happened to me. Many, many thanks for *all* you have done for me.

Adams had observed life in universities, industry, government, the military services, and in the conquered countries of Japan and Germany, but his broad experience had not diminished his naturally buoyant and forward-looking nature. Warm, sympathetic, and compassionate, he nevertheless showed no shallow optimism and no trace of sentimentality. He was pragmatic and practical and exemplified the ethic of hard work and its suitable reward. Unlike his Puritan forebears, he had an enormous zest for life and also believed in relaxation and enjoyment when he put his work aside. His friends at the symposium recognized these traits in him; they all admired his intellectual brilliance and tough-mindedness, but each cherished personal memories of Adams as a complete human being. Thus the

Illinois University Archives

With the book of letters from his students at retirement (1954).

Illinois University Archives

At his retirement dinner (1954).

tone of the symposium, like Adams himself, was cheerful, unsentimental, and informal.

Adams's first Ph.D., Ernest H. Volwiler, president of Abbott Laboratories, opened the symposium and presented a comprehensive picture of the Chief's career.[3]

A series of research papers followed, given by five of Adams's former students and covering a range of organic chemistry as broad as Adams's own studies. Wallace Brode of the National Bureau of Standards discussed the steric effects on absorption spectra and protein absorption of dyes, an outgrowth of his Ph.D. work with Adams. John R. Johnson of Cornell described chemistry of a mold metabolite, gliotoxin; with Adams he had studied organoarsenic compounds as medicinals. Samuel McElvain of Wisconsin, who had investigated local anesthetics at Illinois, presented a paper on the structural chemistry of organic compounds in catnip. Ralph L. Shriner of Iowa, who had been one of the first chemists to work on the development of the Adams platinum oxide catalyst, spoke on flavylium salts. The last speaker was Wendell M. Stanley of the Virus Laboratory at Berkeley, whose student researches had been on the synthesis and testing of compounds for anti-leprosy activity and on optical activity due to steric hindrance to rotation. His symposium lecture described some of his chemical studies on viruses, for which he had received the Nobel Prize.

A dinner in the Illinois Union concluded the meeting, which was enlivened by an irreverent skit about Adams written by some of his students of the late 1930s. The skit burlesqued incidents of Adams's career, and the group, including the guest of honor, enjoyed it to the utmost.

Adams thanked his friends with his usual informality, aplomb, and lack of flowery rhetoric. He obviously enjoyed the symposium as much as anyone and was pleased at the number of those attending. Many years of important activities lay ahead of him, and his view, as always, was toward the future. A picture taken of Adams as he acknowledged the tribute of the crowd captured to a remarkable degree the vitality, the look toward the future, and the inspirational quality that were part of his personality.

What made Adams different from thousands of other chemistry professors in the country? What were the family background, the educational and other factors that shaped his philosophy of undergraduate and graduate training? What were the social and economic factors that made his career possible, and to what degree was he a creator as well as a creature of his times? What is the real significance of his own scientific work, and how did he become a scientific statesman on a world scale? What were the details of his varied public services for science and the national interest?

This study attempts to answer these and many other questions about Adams and his times. The ultimate riddle of human personality cannot be explained beyond a certain limit, but the remarkable nature of Adams and his career deserves presentation. This is true both from the standpoint of organic chemistry as a vigorously growing field and from the broader aspects of Adams's work beyond his own research.

Adams is a prime example of Samuel Johnson's definition of genius: "A mind of large general powers, accidentally determined to some particular direction."

Literature Cited

1. Now in the Roger Adams Archive (hereafter RAA) in the Illinois Archives in the university library at Urbana, oversize folio 3. Hereafter quotations from RAA will have the box number and folder title, in that order; the archive contains 70 boxes and an estimated 50,000 documents.
2. The quotations are from the letters of R. M. Joyce and W. E. Hanford, respectively.
3. The symposium was published as *The Roger Adams Symposium*, preface by C. S. Marvel, Wiley, New York, 1955.

Early Years and College

When Roger Adams was born in Boston in 1889, the son of Austin and Lydia Adams, probably no one in his family imagined that he would become the outstanding organic chemist of his time, with a national and international influence on the development of chemistry and science in general.

Scientific research and teaching were not well-established professions in America in 1889. Although Ira Remsen organized the first substantial graduate school in chemistry at Johns Hopkins in 1876, similar, smaller programs at Harvard, Yale, and elsewhere were still relatively unknown. Because college and university enrollments were still small, the opportunities in college teaching as a profession were very limited, even in the well-established fields.[1]

If his older relatives had predicted the future for Roger, it would probably have been a career in one of the established New England professions—medicine, law, business, or perhaps the ministry or publishing—because the members of his immediate family were well-educated and well-situated in Boston business, and his New England background was unusual.

On his father's side Roger was a direct descendant of the famous Adams family that produced two presidents of the United States as well as several historians and scholars, including Henry, Brooks, and Charles Francis Adams, all grandsons of President John Quincy Adams. Although there is no evidence that this distinguished connection had much influence on the Roger Adams family, they certainly were conscious of it. Roger himself was not greatly interested in genealogy,[2] but membership in a family that had produced many distinguished people, in addition to the President John Adams line, deserves exploration.

The first American Adams of this family was one Henry Adams, described as a maltster and farmer, who arrived in Boston in 1632 or 1633 with a wife, eight sons, and one daughter. He probably came from Braintree in Essex, England. Some believe he was of Welsh ancestry and have furnished a Welsh genealogy for him.[3] Henry Adams settled in Braintree, Massachusetts, with his son Joseph (1620–94).

The latter's son Joseph had eleven children by three wives, and one of these, Deacon John Adams, was the father of President John Adams. Deacon Adams's brother was the Reverend Joseph Adams (1638–1733), the direct progenitor of Austin Winslow Adams, Roger's father.

The Reverend Joseph Adams was born in Braintree and graduated from Harvard College in 1710, receiving his A.M. in 1713.[4] He was ordained in 1715 and settled in Newington, New Hampshire, a small town near Dover, where he was pastor for sixty-six years; his church is still standing, the oldest church building in the state. His friend the Reverend Jeremy Belknap (1744–98) of Dover called him "the Bishop of Newington."

The Reverend Joseph was a leading figure in his region; he assisted in obtaining a charter for Dartmouth College in 1769 and was a founding proprietor of the towns of Rochester and Barnstead in southeastern New Hampshire. As a child John Adams (later president) visited his uncle the Reverend Joseph at Newington and again in Joseph's eighty-second year in 1770, finding him "as hearty and alert as ever." Two of Reverend Joseph's sons, (Doctor) Joseph and Ebeneezer, graduated from Harvard in 1745 and 1747.[4] Joseph was the last of the direct ancestors of Roger Adams to attend college until Roger himself.

Doctor Joseph (1723–1801), who took his A.M. at Harvard in 1748 with a thesis on medicine as a science, became a physician. He lived in Newington and Barnstead in his father's farmhouse, practicing medicine as much as his poor health permitted. John Adams's diary in 1771 says, "at Tiltons in Portsmouth I met with my Cousin [Doctor] Joseph Adams, whose Face I was once as glad to see as an Angel. The Sight of him gave me a new feeling. When he was at Colledge, and used to come to Braintree with his Brother Ebeneezer, how I used to love him. He is broken to Pieces with Rheumatism and Gout now."[4] Thus the Braintree Adams family and the New Hampshire branch maintained contact until at least near the end of the eighteenth century. Whether it extended to the days of Austin Winslow Adams is not known.

The three Adamses intervening between Doctor Joseph and Austin lived in small towns around Pittsfield and Barnstead[6] in the foothills of the White Mountains. They were craftsmen and farmers[7]; some of them probably worked in the Pittsfield cotton mill where average wages might approach $15–$20 per month.[8]

Austin's father, Deacon William Clark Adams (1813–93), was born in Barnstead, one of four children, and he worked in Lowell and Boston for some years as a carpenter and machinist. A memorial booklet described his wife, Elizabeth Wallace Taylor Adams, as active, of high purpose, with good sense and an instinct of success, self-reliant, enthusiastic over excellence, and not given to indirections, all sturdy qualities of her Scotch–Irish ancestry. Allowing for some exaggeration in an obituary, this list of qualities of his paternal grandmother resembles Roger Adams's own personality.

Deacon William returned to New Hampshire, where his son Austin Winslow Adams was born in 1845 in the Barnstead home. William's second son, also named

William,[3] remained near Pittsfield, and his children were the New Hampshire first cousins whom Roger liked to visit when he was young. Adams's relatives were numerous and congenial and it is notable that, at least among Roger's direct forebears, none migrated from their home territory to the Midwest as those states became settled.

Another of the many linear descendants of Henry Adams of Braintree was Herbert Baxter Adams (1850–1901), one of the original faculty members at Johns Hopkins. He trained many of the leading American historians of the next generation in his famous seminar; Woodrow Wilson was one of his students.[9]

On his mother's side Roger also descended from early colonial settlers in Massachusetts. His grandmother, Melissa Hastings Curtis (1811–83), and his grandfather, Ezra Carter Curtis (1812–62), were in the sixth American generation of their large families. Roger's mother Lydia, born in Leominster, Massachusetts, in 1843, was one of eleven children, and her numerous relatives were hard working and prosperous, having wide-ranging family connections with prominent business, professional, political and literary figures. Many of the relatives lived in the handsome suburbs south of Boston—Newton, Milton, and the Highlands—which they helped to develop.[10]

Austin Adams, with this tradition of learning in his background, received a good classical education—while some of the village youth were going into the "cloth room" and the "weave room" of the mills.[11] At nineteen he was examined by the school supervising committee and certified to teach in the rural schools of Barnstead. Not many years later, in 1872, he set out for Boston, as so many backcountrymen did.[12]

Oliver Wendell Holmes had called Boston the hub of the universe, and the intellectual world of New England centered about it.[13] There young Austin had the chance to draw closer to the heritage of his Adams family and to the history of the republic, to walk in the footsteps of New England's gifted men and women, and to browse in well-stocked bookstores. There, too, he had the chance for a better life, a chance to rise above the meager pay of the country schoolmaster. His scholarly tastes, his position, and his family were to keep him in Boston and Cambridge for the rest of his life.

Boston in the 1870s was a growing, flourishing metropolis, one of the busiest ports in the country and a rail center whose lines carried raw materials and factory goods—shoes, textiles, books, and other products—from the industrial city. A center of banking and commerce, its downtown blocks around State Street had been devastated by the Great Fire of 1872, yet in only a few years rebuilding was almost complete. Over all spread the mantle of Boston's culture and history: libraries, museums, and churches were growing, as well as its education centers, Boston University, Boston Tech, and, across the river in Cambridge, Harvard University.

Austin Adams took his first position in Boston as freight clerk with the Old Colony and Newport Railway. He continued in this company all his life, rising to clerk in 1875, to chief clerk in 1879, and to treasurer of the Old Colony Steamboat Company in 1894. In 1879 and 1880 he roomed in Quincy, south of Boston on the railroad line and the historic seat of the Adams family. (Quincy had originally been part of Braintree.)

The Old Colony Railroad Corporation, a Massachusetts institution, was chartered in 1844 to provide service between Boston and the Old Colony, Plymouth. It expanded steadily toward Cape Cod and Rhode Island and in 1854 added the Fall River line with a steamer connection between Boston and New York. By 1880 this prosperous Massachusetts corporation possessed undivided control of transportation in the southeastern corner of the state.[14] During the period of railroad expansion after the Civil War, total investments increased rapidly, fivefold between 1870 and 1890, for example, and wages were well above those of the average American.[15] In comparison with the New Hampshire schoolteacher's average monthly salary of $16.42 plus board (and school did not keep throughout the year), the Nashua and Lowell Railroad paid the stationmaster at Nashua $34 a month.[16] In 1888 the average wage of the rail employee was $572 per year. Austin Adams had made a change for the better.

On September 22, 1880, Austin married Lydia Hastings Curtis of Jamaica Plain, at the Perkins Street home of her uncle Nelson Curtis in that pleasant suburb of Boston.[11] Lydia had graduated from Salem Normal School in 1862 and was certified as a teacher; where she taught before she married is not known. She was thirty-seven years old at the time of her wedding, two years older than Austin, and family tradition says that she had been engaged to a soldier killed in the Civil War.[17] That conflict, so hotly debated by Boston's citizens, took an exceedingly high toll of Massachusetts men in the regiments raised by Governor John Andrew during the war. Austin and his wife made their home at 38 Worcester Street in Boston's South End for almost twenty years.

The Victorian South End was then one of the most attractive residential sections of the city. Historic Boston was founded on a hilly peninsula, its narrow neck dividing Boston Harbor from the mouth of the Charles River. Here were the historical landmarks of old Boston, the great dome of Bulfinch's State House, Boston Common, the business district, and the wharves. West and south of the old city, between it and Cambridge, lay the Back Bay, the tidal basin of the Charles, whose waters were gradually filled in over many decades. In 1850 the available filled section was opened for settlement, and a handsome residential development was laid out, with thoroughfares—Tremont, Washington, and others—running to the suburbs and crossed by streets named for Massachusetts cities and towns. Intended for the prosperous Bostonian, comfortable red-brick homes with mansard roofs and high front stoops edged by diminutive green lawns lined the pleasant streets and shady parks. Worcester Street lay close to the Roxbury line.[18]

Four children were born to Austin and Lydia Adams: Mary (1882), Emily (1883), Charlotte Hastings (1885), and Roger (January 2, 1889).[3] The Adamses must have found these early years of their marriage difficult, for, along with the births

Lucile Adams Brink

Austin Winslow Adams (1845–1916).

Lucile Adams Brink

Lydia Curtis Adams (1843–1921).

Lucile Adams Brink

The Adams Children:
rear, Charlotte, Mary, Emily; front, Roger.

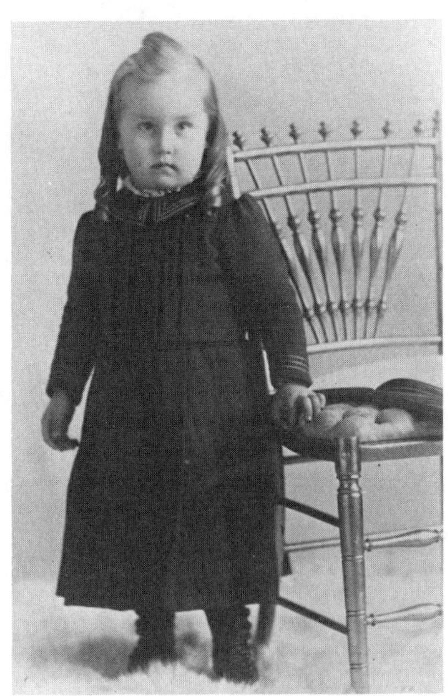

Lucile Adams Brink

Roger Adams (ca. 1891).

of the children, they suffered family deaths. Uncle Nelson Curtis died in 1882; Lydia's mother, brother Samuel, and sister Sarah died within the next two years; and Austin's mother, Elizabeth Adams, died in 1888.

Pictures of this period from the family album show that Austin Adams was a rather short, pleasant, full-bearded man with a high forehead and quiet appearance; Lydia was a tall, striking woman with a long face and firm, handsome features, beautifully and fashionably dressed in the full-skirted, tightly fitted jacketed suit of the day. The many carefully posed, stylish photographs of the children show the girls "Birdie" (Mary), "Daisy" (Emily), and Charlotte in starched white dresses, black stockings, and buttoned kid boots, while little Roger, the family pet with golden hair and blue eyes, stood solemnly in his black stockings and boots, his stiff dark dress topped with a lace collar and his thick hair in long, carefully rolled curls. His downturned mouth, easily recognizable all his life, is clear in his childhood pictures.[19]

Adams maintained a comfortably well-to-do household.[20] Ellen, the cook, and Sarah, the maid, ruled in the high basement kitchen, and there was always a nurse, Nellie or Lamb, for the youngest child, or Roger's governess, Mary, who was in charge of the children's upstairs nursery. A guest room was available for Cousin Fanny, Aunty Em, or Cousin Charlotte Hastings Whipple, while on the main floor spacious parlors with sunny bay windows, dining room, and library accommodated entertaining and study.

The family were devoted members of the congregation of the New Old South Church built in 1877. Its Italian Gothic style and lofty bell tower anchored one of the corners of splendid Copley Square, along with the (old) Art Museum, Trinity Church (where Phillips Brooks preached), and later the Boston Public Library (1884), all easily accessible from Worcester Street by the horse trams. Years later the Reverend George A. Gordon, called to New Old South in 1884, was to remember Lydia's "great nature and her loyal support," this when Gordon's freer interpretation of the traditional New England Congregational theology caused friction among the worshipers. Austin Adams he recalled as the "ablest intellect" in the church; an essay of his was "the equal of Dr. Bushnell," the noted theologian of Hartford, Connecticut.[21] The family attended church and Sunday school and held evening prayers at home. An early example of Roger's determination was his walk to Sunday school through the terrible blizzard of 1898 when he was nine years old; he was the only pupil present.

The children all attended the private Everett Grammar School on Northampton Street a few blocks away; Ellen escorted them there daily on the horse cars. Roger's long curls were cut in 1894 when he entered school; across the Common on Brimmer Street in old Boston, Sammy Eliot Morison and his brother also suffered from the hated velvet dresses and suits with lace collars and the long curls made fashionable by Frances Hodgson Burnett's *Little Lord Fauntleroy* in 1886.[22] The girls enjoyed drawing, painting, and singing lessons, and Madam Blackstine of Paris was the "French mistress for the Adams girls."

In the fall of 1899 Roger entered the seventh grade in the public Boston Latin School in the South End. Besides collecting weekly cards for good conduct, he became attached to public declamation and on many evening occasions recited addresses by famous orators—McKinley, Everett, Barré—and represented his class in the Annual Prize Declamation of the Boston Latin School in June 1900, when he declaimed "A Monument to Shakespere" by Hugo, all good training for the uncountable speeches in his future.

The house on Worcester Street, filled with four lively, mischievous youngsters, rang with laughter and activity; the Reverend Gordon remembered well their mirth in the Boston home and later in Cambridge when he used to call on them. There were so many things for the children to see—Sarah lighting the gas chandeliers in the parlors and the city lamplighter out of doors; coaches and fire horses dashing down the street and, in the winter, sleighs and working pungs; milk wagons and ice wagons and Papa's trains puffing at the depot. Horsecars plodded by until the electric poles went up in front of the house in 1892, and electric streetcars followed soon after all over Boston.

Days were filled with school and play; summer jaunts to Curtis relatives' country homes in the cooler suburbs of Milton and Newton; summer visits to the old homesteads in Pittsfield and Barnstead where the Adams cousins shared farm chores, boating, and excursions, trips to Boston and the Public Garden on the streetcars. Fourth of July celebrations crackled with speeches and fireworks, and on Independence Day in 1890 Papa sent up two tissue paper balloons five feet in diameter! Bicycling was all the rage after the safety bicycle came out in the nineties; the girls had theirs, and Roger rode with Papa until he proudly showed off his own in a picture taken when he was eight. On a splendid fall outing in 1897,"Old Peppersass," that doughty old engine with upright boiler and stack like a pepper pot, hauled the whole family up the cog railway to "Cloudland," Mount Washington's summit.[23]

Amateur dramatics, church socials, and entertaining were balanced by quiet scholarly achievement. As Austin Adams worked on literary papers, the Boston bells were tolling for the passing of New England's great men: Longfellow, 1882; James Russell Lowell, 1891; Whittier, 1892; Phillips Brooks, 1893; Dr. Holmes, 1894; and Francis Parkman, 1893. Their pictures were cherished in the Adams house. A well-preserved clipping from the *Boston Transcript* in 1894 told of the long-awaited selection of the fifty-three "celebrated sons of Massachusetts" to be inscribed around the base of the dome in the new chamber of the House of Representatives, names known to every Boston family and resounding to them: John Adams and John Quincy Adams, Emerson, Hawthorne, Carver, Endicott, Morse, Webster, Garrison, Sumner, and Agassiz.[24]

In their own family (as if the Adams name was not enough in Boston), Edwin Percy Whipple (1819–86), essayist, lecturer, and banker of Boston, was the husband of Cousin Charlotte Hastings, Lydia's first cousin and dear to the family. Annie Adams (1834–1915), distantly related, married James T. Fields (1817–81), author and publisher of the famous publishing house Ticknor and Fields. Mrs. Fields was a much-loved literary hostess of Boston and a warm friend of Charlotte Whipple; there was a long association between these two families.[25]

During these years of Boston residence Austin continued with the Old Colony Railroad system. With 600 miles of track, it was one of the best-managed and most gilt-edged investments in railroading. The Fall River line with its famous boat train from Boston to Fall River and its luxurious steamers on to New York were national institutions. Vacationers to the South Shore and Cape Cod, with connections to the islands of Nantucket and Martha's Vineyard, filled the yellow cars of the passenger lines. Nevertheless, in 1894 the Old Colony succumbed to the spreading embrace of the powerful New York, New Haven, and Hartford Railroad and was taken over. The profitable railroading years near the turn of the century resulted in a wave of station building: elaborate and imposing stations went up across the country, and in 1898 Boston's new and stately South Station opened as the combined terminal for the Boston and Albany and the New Haven systems; for years it remained the largest and busiest station in the country.[26] At that time Adams became treasurer of the Boston Terminal Company (organized in 1898) with his office in "our station," of which he was very proud, and he remained in this position until his death.

At the turn of the century the Adams family made an important change, leaving Boston and moving to Cambridge. Adams's business office was readily accessible from Cambridge; electric streetcars had run between the two cities since the nineties. The children were growing up, and their parents, in a decision reflecting the scholarly attitude of the home, faced the prospect of putting four young people through college; a move across the river would enable some of the students to live at home. In addition, the beautiful South End was changing. After 1870 the Commonwealth completed the filling of Back Bay, and the new district—west of the Public Garden along Beacon Street and park-centered Commonwealth Avenue—became the fashionable residential area of Boston. With the improvement in interurban transportation, the laboring population, predominantly Irish in the South End, began to move out of the densely populated old city. Factories and tenements pressing into the pretty streets of the South End hastened the changes; the pleasant homes became lodging houses, and the churches geared up for settlement work.[27]

In Cambridge in 1900 the Adams family settled at 8 Wendell Street,[28] a narrow street faced by tall wooden houses, just off Massachusetts Avenue where cars ran directly to Boston. A fifteen minute walk brought one to prestigious Cambridge Latin School on Trowbridge Street and Harvard and Radcliffe Colleges. Mary, Charlotte, and Roger attended the Latin School, and Emily entered Radcliffe in 1900.

Mary, who had never been healthy,[17] spent one year at Cambridge Latin; she was a good student and entered Smith College in 1902. She graduated in 1907 (having lost a year because of illness) with major studies in French, literature, and history.[29] She lived with the family as a boarder and taught school, perhaps in girls' schools or at the New Old South Church. In 1924, leaving Cambridge, she entered the Home of Truth in Boston, a religious group with whom she lived until her death in 1928. In a touching letter of condolence to Emily, her minister, the Reverend George Gordon, wrote of Mary's splendid and loving nature and her teaching in the church.[21]

Emily entered Radcliffe in 1900, and she and Charlotte, who entered in 1902, lived at home and enjoyed lively years.[30] The Radcliffe students were not allowed to attend Harvard classes, although members of the Harvard faculty conducted instruction at the Radcliffe Yard on Garden Street. At Radcliffe Emily was an officer of the Emmanuel Club (the dramatic society) and starred in class basketball. Noted for her "histrionic talent," she spoke at commencement in 1904 when ninety-three women graduated, including Helen Keller. After graduating, Emily taught in private girls schools, teaching college preparatory subjects (mathematics and physics), basketball, and dramatics, and later teaching dance at Abbot Academy in Andover, Massachusetts. Her emotional nature and her interest in spiritual healing led her to join the Home of Truth also, and she remained a member of this group for the rest of her life. In Emily's last years she was partially blind, and both Charlotte and Roger supplied funds to keep her in rest homes. She was very proud of her brother, and Roger, who wrote long descriptive letters of his travels to her, spent much time and effort in making her comfortable.[31] A note to Radcliffe reported Emily's death in 1970.

Charlotte Hastings Adams, who graduated cum laude from Radcliffe in 1906, was a gifted, spirited, and admired member of her class, busy with many extracurricular activities: in her own words, "much dramatics, Glee Club, Mandolin Club, Class basketball team, Emmanuel club, etc." In her senior year she gleefully wrote the Class Prophecy, in which she gave full scope to her sense of humor and of the ridiculous. Tall, dark, and willowy, she was listed as an "athletic champion" and, like her sister, spoke at commencement, presenting "The Comedy Part."

Charlotte taught for five years at the Gilman School in Cambridge and the Winsor School in Boston, giving declamation, reading, and music; during this time she lived with the family in Cambridge as a boarder. Adventuresome as always, in 1913 she made a determined fling at acting, which she described as "five years of intensive character study and portrayal in various stock companies all over the country from old Castle Square to the Burns Theatre in Colorado Springs." Castle Square Theater on Tremont Street was a Boston institution in its day. Operated by its manager, producer, stage director, and star actor, John Craig, it presented the year-round varied repertory of a "high-class stock company"; Charlotte played one season in *The Product of the Mill.*[32] Her appearances in various repertory companies took her out west, where, with characteristic verve, she rode up the new Pike's Peak auto road only two years after its opening. She thrilled to the enthusiastic audiences on this western tour, and she shared her experiences with her family at home by letters and pictures. Her father, poking fun at himself for demonstrating his pride, took her pictures to his office to show to everyone.[33] It was a disappointment to her that acting was not to be her career.

In 1918 her stage career ended and, her money running out, Charlotte moved in with a classmate, Helen Wheeler, and took an intensive ten-week secretarial course in New York. Employed by Columbia University, she rose rapidly, in a variety of jobs, to recording secretary of the Medical School. She demonstrated great skill in managing personnel and in directing the complex duties of university offices. Charlotte was fond of the theater, traveled widely, and enjoyed sketching and painting. At one time she did pencil portraits of Roger Adams, his wife, and their daughter.[34]

Charlotte never married;[35] she retired in 1943 to live in Pittsfield, Massachusetts, where she had played in stock theater, and died in 1947. Her many friends remembered her as a talented, warm-hearted, and intrepid woman, and Columbia University paid her sincere tribute: at her death the flag at Columbia flew at half-mast.

Roger Adams was eleven years old when the family moved to Cambridge and he entered the eighth grade at Cambridge Latin School.[36] There he took a classical college preparatory course including Greek and Latin in which he did well. Otherwise his grades were fair and he stayed within the top half of the class. He was not busy in extracurricular activities at school; none, at least, shows in his class book, *The Cambridge Review, June 1905,* where the "alphabet of celebrities" rhymed:

> R's for R. Adams, '05's smallest son, What he lacks in his height he makes up in his fun.

He apparently had fun in Cambridge and, best of all, at Camp Wildamere in Maine, which offered mountain climbing, water sports, tennis, canoe trips, and, most exciting, "coaching trips," touring and camping out through the New Hampshire mountains. Twice they climbed Mount Washington, and in a long day in 1904 Roger was the second camper to reach the top. In 1905 he attended the Pine Island Summer Camp; there he "covered first base" in the ball games.

His school hours were undoubtedly long and arduous as a junior and senior, because he entered Harvard on the "Old Plan," accumulating points by taking the special Harvard entrance examination for two years at Sever Hall and entering with the traditional sextet of classics (advanced Greek and Latin), mathematics (plane geometry and algebra), modern languages (French), English, history, and laboratory science (chemistry). It was the last year that Harvard recognized only its own admission examinations; in 1906 the university accepted the College Entrance Examination Board standards.[37] Roger's admission won Austin Adams's gentle accolade: "Bene, mi fili."

The four years that Roger attended Harvard College[38] were President Charles William Eliot's last four years. During his distinguished career from 1869 to 1909,

Eliot built Harvard into a great university with an eminent faculty offering an impressive variety of courses and imbued with lofty ideals of teaching and research, graduate professional schools of high caliber, and, for the student, unprecedented freedom and flexibility in choosing his own college education by the elective system of courses. Roger's entrance coincided with the zenith of free electives, when a student might obtain his degree by blithely sampling rich offerings of the whole college without formulating a coherent program, or, on the other hand, might overspecialize in one field.

Roger entered Harvard at sixteen, a stripling of 5 feet, 5½ inches, weighing 125 pounds. At that time he was sensitive about his height. Perhaps the teasing jingle in the Latin School yearbook, perhaps his tall athletic sisters starring in basketball, or perhaps his freshmen classmates, usually one to three years older than he, prompted him to send away for a program of stretching exercises. Although recorded figures did show him to be below average in all physical measurements except leg and arm muscles (thanks to mountain climing and tennis), he was too level-headed and practical to be long disturbed.

In Roger's class at Harvard eighteen Cambridge Latin boys entered together, the third largest group next to Boston Latin and Groton. Friends in that number were an advantage when freshmen had no special dormitories and, unless they crashed society parties or clubs, had only a few mixers and party invitations before they were abandoned to their own devices. Roger occasionally assisted at church socials in the New Old South Church, where he remained a faithful member.

He started a heavy freshman program of five courses: Chemistry 1, the required English A, a double course in beginning German, and mathematics. His grades, B in chemistry, C's in German, and D's in English and mathematics, were poor but acceptable then, when Harvard "was the easiest [college] to stay in" and C's were "gentlemanly."

It is not now clear what influenced Roger to major in chemistry, but perhaps his first college course aroused his enthusiasm.[17] Chemistry 1 had been given since 1870 when President Eliot, himself a chemist and notable teacher responsible for the first student laboratory instruction at MIT, initiated a new course with both lectures and laboratory work at Harvard. This proved wonderfully successful under C. Loring Jackson, who taught it from 1870 to 1912, and its enrollment swelled from 40 students to close to 400. Jackson attributed much of its success to material on the "chemistry of common life," fragments of applied science illustrating many processes of chemical industries offering jobs to graduates, such as baking, iron, steel, and glass production, "the theory of manures," bleaching, respiration, and photography. Jackson's elegant and humorous style, his pride in the course, and abundant lecture demonstrations together with laboratory instruction no doubt contributed much. Actual laboratory experience for undergraduates was unusual in those days, and the opportunity for personal investigation and preparations, even in the jammed cellars of Boylston Hall, drew interested students.[39] Roger had also passed chemistry with a good grade in his sophomore year at Cambridge Latin; his teacher there was J. I. Phinney. (A.B. 1892, Yale), a graduate student for a time at Harvard with published research done under Jackson.[40]

Whatever the impetus, in his sophomore year he was obviously committed to chemistry and completed an arduous program of six college courses: four full courses and two half courses each semester. Into this one year he crammed full courses in qualitative and quantitative analysis (both with heavy laboratory schedules) under Charles R. Sanger and Gregory P. Baxter, respectively, physics (with laboratory), German literature, and half courses in organic chemistry (with no laboratory) given by Henry A. Torrey (assisted by J. Enrique Zanetti, later to go to Columbia), historical and elementary physical chemistry by Baxter and Torrey, advanced algebra, and English composition (the hated sophomoric themes). He did well to collect four C's and two D's, the D's being in English, German, and historical chemistry. The Harvard grading system was based on four full courses, thus D's in two half courses counted as one D for the year.

In this decade when college expenses mounted, Austin Adams suffered reverses in several panics, especially the panic of 1907[17] caused by a period of financial stringency after four prosperous years. Investments in railroads, and particularly those relating to the New York, New Haven, and Hartford, were critical. A prosperous system paying dividends of eight percent when it took over the Old Colony in 1894, it had been greatly affected by the manipulations of its president, Charles S. Mellen, who recklessly overextended its expansion. Passed dividends and stock plunges caused loss and dismay to many Boston investors.[41] Although Adams was never unemployed in these years, a drop in income was particularly unwelcome.

Roger, who seemed to need no spurring on, probably worked one summer in silver mining,[17] perhaps in the vacation between sophomore and junior years when he took no summer school classes. At the end of the summer he and a friend signed on a cattle boat to work their passage to England where they had a week free in London—a popular exploit of the times.

Clearly Roger had decided to complete his undergraduate work in three years, a decision not uncommon at Harvard in this period. An ambitious student, by "anticipating" courses prior to entering college or by taking extra courses during the year and attending summer school, could finish early yet be listed with his four-year entrance class. Roger completed five and one-half courses in 1907–8: advanced organic chemistry under Torrey, and three in metallurgy, mineralogy, and mining (with laboratory work in fire assaying), all allowed in a chemistry major, along with two electives, French literature and economics. His improved record of A's (in organic chemistry, mineralogy, and French), and B's (in metallurgy and mining), with D in economics,[42] placed him in Group I, and he received a "Detur," as Harvard termed its book prize, and an honorary John Harvard Scholarship for scholastic merit in the Harvard Prize Awards ceremony in December 1907. With summer school classes in advanced geometry and advanced physics

Lucile Adams Brink

Adams as a graduate student (ca. 1912).

under Percy W. Bridgman, a Nobel Prize winner in 1946 for work in high pressures, he completed his requirements for graduation.

This intensive three-year push showed that Roger found his interest in chemistry in his freshman year and pursued it tenaciously. He was under some financial pressure at home because his rapid progress had actually been more costly annually, as Harvard had charged extra tuition for more than four courses per year since 1902, but by the fall of 1908 he was free to take a leave of absence, or, as many did, to enter the Graduate School of Arts and Sciences. In his class (1909), 44 members out of the 496 who graduated finished in three years; early completion was already a fading trend.

In his fourth year, actually his first year of graduate study, he took four half courses, one of them being his first class with Theodore William Richards, who was then engaged in the very accurate determinations of atomic weights of elements, the work that was to bring him the Nobel Prize in chemistry in 1914. This was a graduate class in physical chemistry, lectures without laboratory (which he took the following summer); his grade, B. He also took gas analysis with difficult laboratory work under Baxter, C; biological chemistry with L. J. Henderson, B; reactions of organic chemistry given by Torrey, A; and research with Torrey, B. It was a striking record of more than twenty-one courses in four strenuous years completed by a young man of twenty.

President Eliot announced his resignation in November 1908 after forty years of untiring effort in transforming Harvard to a "university in the largest sense," and Abbott Lawrence Lowell, professor of government, assumed the president's duties in May 1909, upon Eliot's actual retirement. The commencement activities rolled on at the Yard for ten days with Baccalaureate, two Harvard–Yale baseball games, and all the appropriate Class Day festivities. Roger belonged to the Boylston Chemical Club founded for undergraduates by Richards in 1885; this, Group I, his Detur, and his John Harvard Scholarship were all that his class album listed. In later years James B. Conant remembered the long hours of laboratory sessions and noted that few chemistry majors at Harvard could afford the time for extracurricular activities such as athletics or editorial positions in undergraduate publications.[43]

Other classmates who had been active in their college years received their diplomas from President Lowell on June 30, 1909, and many were to be well known in later life: George H. Edgell of Harvard's School of Architecture; William C. Graustein, mathematician; Dexter Perkins, historian; Lee Simonson of the theater; Hans von Kaltenborn, journalist and dramatic news broadcaster; Theodore Roosevelt, Jr., explorer and military general; Henry Beston Sheahan, natural historian and writer; Ernst Franz Hanfstaengle, who rose with Hitler in Germany but escaped to England before the war. No one could have foreseen that an account of the career and honors of little-known Roger Adams would fill page after page of the *Fiftieth Anniversary Report* of his class in a remarkable record of an active professional life.[44]

The family moved in 1910 from Wendell Street to 7 Riedesel Avenue, a modest 2½-story clapboard and shingle dwelling on a quiet side street off Brattle Street and the famous "Tory Row" of Cambridge. Here Roger boarded during the rest of his days at Harvard, and this was the last home where all the family was together.

A short walk down Brattle Street took Roger to Radcliffe, where he held an appointment at $650 a year for 1909–10, assisting in one of the chemistry courses given by Harvard faculty members in the small laboratories erected in the Radcliffe Yard during the 1890s. His zeal for research and study in advanced topics in organic chemistry under young Dr. Torrey resulted in straight A's. Henry A. Torrey (1871–1910), a graduate of the University of Vermont, had taken his Ph.D. at Harvard in 1897 as H. B. Hill's student. After teaching at Vermont, he joined the Harvard faculty in 1903 where he worked on quinones, azo dyes, and alkali-insoluble phenols. Torrey won regard as "an inspiring teacher and investigator of great promise",[45] and his premature death in March 1910 ended the three-year association with Adams. Roger completed Torrey's teaching obligations to Radcliffe, and later, with the help of Jackson and G. S. Forbes, he finished his problem on alkali-insoluble phenols. At commencement in 1910 he received his A.M. degree.

He carried out additional research with Latham Clark, a junior faculty member, on the synthesis of aliphatic hydrocarbons, and also with Richards in physical measurements. During the spring of 1911 he was able to foresee the completion of the work for his doctorate by the middle of the following year. With characteristic directness he asked and received permission from Richards to take his examinations in inorganic and analytical chemistry in June 1911: "This will greatly facilitate the work of preparation." In his last year of graduate school, 1911–12, he held an Austin Teaching Fellowship.

He submitted his dissertation for the Ph.D. on May 1, 1912. It consisted of three parts: "I–A study of the Solubilities in Aqueous Alkalis of Various Hydrazones of Certain Aromatic Ortho-Hydroxyaldehydes and Ketones; II–Nonanes; III–A New Bottling Apparatus." Each section bore a sincere expression of appreciation to his research directors, and he paid handsome tribute to Torrey. His work with Torrey gave him good experience in the synthesis of aromatic compounds, which he was to find valuable in the future, and from the research on hydrocarbons with Clark, "he learned first hand (among other things) the practical problems and limitations associated with the then existing methods of catalytic hydrogenation. In these formative years Adams took great pleasure in experimental work with organic compounds, a zest which would later be imparted to many coworkers."[46] For Roger the laboratory work with Richards, which involved exacting manipulations in the determination of the atomic weight of uranium, lacked the excitement of organic chemistry.

Twenty-nine men formed the roster of graduate students in chemistry in 1911–12, and Adams made acquaintances at all levels through course work and assisting. Among them were Elmer Keiser Bolton ("Keis" to his associates), a special friend of Roger's and later director of the Chemical Department at Du Pont; Webster N. Jones, later of Goodrich and provost of Carnegie Tech; Farrington Daniels, who enjoyed a distinguished career at Wisconsin; Frank C. Whitmore, who would go to Pennsylvania State; James B. Sumner, also one of Torrey's students, who was to be a corecipient of the 1946 Nobel Prize for pioneer work in enzymes at Cornell; and Emil R. Riegel, who moved into industrial chemistry at the University of Buffalo.[47]

That spring Webster Jones and Farrington Daniels, both members of Alpha Chi Sigma at their undergraduate colleges, succeeded in starting the Omicron Chapter at Harvard. Eleven undergraduates and eight graduate students, Adams among them, became initiates when the chemistry faculty approved the fraternity with some reluctance, as the Boylston Chemical Club and the Graduate Chemical Club already existed. The members, including Bolton, Daniels, and James B. Conant (A.B. 1914, Ph.D. 1916, Harvard), at that time an undergraduate enrolled in advanced courses and later president of Harvard University (1933–53), became Adams's life-long friends.[48] There was also the very selective, secret Carbon Club, where flowing conviviality enlivened sharp discussions on research, and Roger added his beer stein inscribed "Adams" to those of other graduate student and faculty members.

For Adams, hard working, fun loving, and gregarious by nature, these years of comradeship and participation in creative thinking and lively inquiry in the exploding field of chemistry and other topics of the times, plus the demonstrated opportunity for a student to work his own way through graduate school, were of fundamental importance. All these activities characterize a graduate student group and must have had an influential part in molding his ideas of university education in America.

Literature Cited

1. Hugh Hawkins, *Pioneer: A History of the Johns Hopkins University, 1874–1889,* Cornell Press, 1960; D. S. Tarbell, "Organic Chemistry: The Past 100 Years," *Chemical and Engineering News,* April 6, 1976, pp. 110–23; Frederick Rudolf, *The American College and University, A History,* Knopf, New York, 1962.
2. RAA, 16, Personal Correspondence 1960–67; Adams wrote April 3, 1967, ". . . I am only casually interested in genealogy. . . ." Adams owned the book cited in the following note, but also wrote that it had some (unspecified) errors.
3. Andrew Napoleon Adams, *A Genealogical History of Henry Adams of Braintree, Mass., and His Descendants; also John Adams of Cambridge, 1632–1897,* Tuttle Company, Rutland, Vt., 1898.
4. C. K. Shipton, ed., *Sibley's Harvard Graduates,* Vol. V, Massachusetts Historical Society, Boston, 1937, p. 502; L. H. Butterfield, ed., *Diary of John Adams,* 4 vols., Harvard Press, Cambridge, 1961; June 30, 1770, *1,* 354, for Reverend Joseph. *Sibley's Harvard*

Graduates, Vol. XI, 1960, p. 513; *Diary of John Adams*, June 24, 1771, *2*, 40, for Doctor Joseph.

5. Jeremy Belknap, *History of New Hampshire*, 3 vols., Boston, 1784–92; reprinted by Arno Press, New York, 1972; Ref. 3. Belknap was Harvard 1762.
6. Lucile Adams Brink papers (hereafter L. A. Brink papers).
7. RAA, 1, Genealogy.
8. E. Charlton, *New Hampshire As It Is*, 3rd ed., Tracy and Company, Claremont, N.H., 1856. Articles on Pittsfield and Barnstead, pp. 349 and 100; on wages in textile mills, pp. 287, 312, 325, 390; on education, p. 482. E. S. Stearns, ed. *Genealogical and Family History of the State of New Hampshire*, 4 vols. Lewis Publishing Co., New York and Chicago, 1908. Vol. 1, p. 185: Adams family; Vol. 2, p. 673: Winslow family.
9. H. W. Bragdon, *Woodrow Wilson, The Academic Years*, Harvard, 1967, passim.
10. V. S. Fulham, *The Fulham Genealogy*, Free Press Printing Co., Burlington, Vt., 1910, pp. 133–34; Lydia N. Buckminster, *The Hastings Memorial*, S. G. Drake, Boston, 1866, pp. 149–50; both are in the L. A. Brink papers and give the genealogy of Lydia Curtis. Lydia's first cousin, Edwin U. Curtis (1861–1922), became mayor of Boston in 1894.
11. RAA, 1, Austin and Lydia Adams. His certification for district no. 3 is dated Barnstead, November 12, 1864. His wedding certificate and invitation are here also.
12. *Boston Directory*, Adams, Sampson and Co., and later Sampson, Davenport and Co., Boston. Volumes 1869 through 1901 give the addresses and business connections of Austin Adams.
13. Van Wyck Brooks, *New England: Indian Summer 1865–1915*, Dutton and Co., New York, 1940. Chapter I, "Dr. Holmes's Boston," pp. 1–22, gives a mellow account of Boston's appeal and its newcomers following the Civil War.
14. Justin Winsor, ed., *The Memorial History of Boston 1630–1880*, 4 vols., Ticknor and Co., Boston, 1881–83; Charles Francis Adams, Jr., "The Canal and Railroad Enterprises of Boston," Vol. 4, pp. 140 ff. This member of the Adams family, historian and railroad executive after the Civil War, was chairman of the Massachusetts Board of Railroad Commissioners from 1872 to 1879. We have found no connection, but it is interesting to speculate that Austin Adams entered the railroad industry with the help of his illustrious distant cousin, and that the years he spent in Quincy linked the two branches of the family.
15. J. F. Stover, *American Railroads*, University of Chicago Press, 1961, pp. 175–79.
16. A. F. Harlow, *Steelways of New England*, Creative Age Press, New York, 1946, p. 360. An entertaining and detailed history.
17. Information received from L. A. Brink.
18. W. H. Kilhan, *Boston After Bulfinch*, Harvard University Press, Cambridge, 1946, Chapter VI, "Expansion," pp. 57 ff. This book is an able and fascinating delineation of Boston's architectural growth.
19. Most of the information about the Adams family in Boston has been pieced together from two sources: (1) Refs. 7 and 11 and RAA 62 and 63, Family Photographs; (2) family materials and photographs in the L. A. Brink papers. Mrs. Adams was a devoted mother and kept many notes on the children's early doings and bright sayings; these scraps illuminate the Adams's home life in Boston in a most charming manner.
20. E. E. Clark, *Clark's Boston Blue Book*, the Elite Private Address, Carriage and Club Directory. Ladies visiting list and shopping guide for West End, South End, Highlands,

South Boston, Charlestown, Jamaica Plain, Dorchester, Brookline, and Cambridge; E. E. Clark, 41 West Street, Boston, 1900. The Adams household at 38 Worcester Street is listed.

21. RAA, 1, Emily Adams Correspondence. George A. Gordon, D.D., in a letter of condolence to Emily on the death of Mary from 83 Longwood Avenue, Brookline, Mass., dated Feb. 9, 1928. Also, Gordon, an eminent member of Boston's nineteenth century constellation of preachers, in his autobiography *My Education and Religion*, Houghton Mifflin Co., Boston and New York, 1925, expresses his views on accepting the pulpit in Boston and his associations there from 1884 to 1925, pp. 255 ff.
22. Samuel Eliot Morison, *One Boy's Boston, 1887–1901*, Houghton Mifflin Co., Boston, 1962. Morison (1887–1976) was two years older than Roger Adams; his delightful book mirrors and suggests many of the experiences that the Adams family must have had in this gay and gaslit era, although it is probable that Roger's family with older parents lived a less active life than the Morisons and Eliots of Beacon Hill. Mrs. Brink remembers her father, Roger Adams, as saying that his best times were in the country. Later Morison, an eminent historian, was a contemporary of Adams at Harvard, receiving his A.B. in 1908 and his Ph.D. in 1912.
23. RAA, 2, Mount Washington.
24. RAA, 1, Prints and Photos; *Boston Transcript*, November 10, 1894.
25. E. W. Emerson, ed., *The Early Years of the Saturday Club, 1855–1870*, Houghton Mifflin Co., Boston, 1918. Chapters "Edwin Percy Whipple" by Bliss Perry, pp. 117–23; "James Thomas Fields," by M. A. DeW. Howe, pp. 376–87. Also RAA, 1, Edwin P. Whipple Correspondence and Charlotte Hastings Whipple Correspondence. Charlotte Hastings Whipple had visited England, and prints of the ancestral village scenes were kept by the family. Also *Dictionary of American Biography*, Annie Adams Fields, James T. Fields, and Edwin P. Whipple.
26. Ref. 16, Harlow, Chapter 10, pp. 215–35, "In the Land of the Pine and the Cranberry Bog," the history of the Old Colony Railroad. Also, E. P. Alexander, *Down at the Depot, American Railroad Stations from 1831–1920*, Clarkson N. Potter, Inc., New York, 1970, p. 219.
27. Ref. 18, Kilham, pp. 62–63. Also R. A. Woods, ed., *The City Wilderness, A Settlement Study by Residents and Associates of the South End House*, Houghton Mifflin Co., Boston, 1898, p. 1, "Introductory," by W. I. Cole.
28. *Cambridge Directory*, W. A. Greenough and Co., Boston. Volumes 1901 through 1927 give the Adamses' residences and occupations.
29. RAA, 1, Mary Adams Correspondence; a copy of her Smith College transcript is here.
30. RAA, 2, Radcliffe, Emily and Charlotte Adams. See also Vita Folders of Emily Adams, 1904, and Charlotte Adams, 1906, in Alumnae Files, Radcliffe College Archives, which contain source material for the Radcliffe alumnae directories of 1928, 1931, 1934, and 1940, and other papers; as well as their class books and reunion books for fifteenth, twenty-fifth, and fiftieth reunions. A note on Emily's death is in the *Radcliffe Quarterly*, 54, No. 4, 31 (1970), and an obituary of Charlotte, written by her classmate and lifelong friend Helen Bridgham Wheeler, is in the *Class of 1906, Fiftieth Anniversary Record, June 1956*. We are also grateful for the reminiscences of William M. Wheeler, the son of Helen Wheeler, who recalled his acquaintance with "Aunt Charlotte," and to the late Theresa Norton Turner, Radcliffe 1906, who, in her nineties, still spoke with affection of the lively and talented Charlotte.

31. RAA, 16, Personal, Family, 1963–67.
32. RAA, 4, 1912; contains *Castle Square Program Magazine*, Vol. II, Nos. 22 and 24, February 5 and 19, 1912.
33. Ref. 6, Austin Adams to Charlotte ("Dear Puss . . . From Pa"), October 17, 1916, tells of his pleasure in her career.
34. RAA, 63, Portrait and Group Photographs, 1889–1971.
35. Ref. 17. Mrs. Brink recalls an intimation that Lydia Adams's trials as a homemaker and mother of four discouraged her daughters from thinking of marriage.
36. RAA, 2, Cambridge Latin School and Mount Washington.
37. S. E. Morison, *Three Centuries of Harvard*, Harvard University Press, Cambridge, 1937. For entrance examinations, pp. 369 ff.; for material about Harvard in President Eliot's years, pp. 323–438; for President Lowell's first years, pp. 439–46.
38. RAA, 3, Harvard, 1905–1912.
39. S. E. Morison, ed., *The Development of Harvard University*, Harvard University Press, Cambridge, 1930. C. L. Jackson and G. P. Baxter, "Chemistry, 1865–1929," pp. 258–76. For Chemistry 1, Parts 1 and 2, pp. 260 ff., by Jackson. Also, G. S. Forbes on C. L. Jackson, *Biograph. Mem. Nat. Acad. Sci. 37*, 112 (1964).
40. D. S. and Ann T. Tarbell, unpublished material on American chemists. There is also a tradition in the Wheeler family (*see* Ref. 30) that Plumer Wheeler, Harvard 1902 in chemistry and the fiancé of Helen Bridgham, Charlotte's classmate, had known young Roger and advised him to take up chemistry; Wheeler found it of great interest. Years later, Mrs. Wheeler was to ask D. S. T. about a student she and her husband had known, "Did you ever hear of him? He took up chemistry. His name is Roger Adams."
41. Ref. 16, Harlow, pp. 331–37.
42. It is ironic that in two subjects, English composition and economics, Roger garnered only D's; in later life the ease and perfection with which he wrote speeches and papers and his acuity in business affairs were astonishing. "Deturs" were established in the 1600s by Edward Hopkins of London and Connecticut for students in "foreign plantations," *Harvard Bulletin*, December 16, 1980, p. 5.
43. J. B. Conant, *My Several Lives*, Harper and Row, New York, 1970, p. 24, hereafter quoted as Conant. Just a few years behind Adams, he gives a perceptive and detailed description of chemistry at Harvard at that time, pp. 20–40.
44. *Harvard College—Class of 1909, Fiftieth Anniversary Report*, Harvard University Press, Cambridge, 1959, pp. 4–8.
45. Ref. 39, Part 3, "The Period 1912–1929," by G. P. Baxter, pp. 269–76.
46. E. J. Corey, in his chapter on Roger Adams in the "*Proceedings of the Robert A. Welch Foundation, Conferences on Chemical Research, XX. American Chemistry—Bicentennial*, W. O. Milligan, ed., Houston, 1976, p. 204.
47. Harvard Archives, UAV.275.5, Dept. of Chemistry Correspondence, boxes for 1910–1911 and 1911–1912; RAA, 5, Harvard, 1909–1911; and *American Men of Science*.
48. RAA, 4, Alpha Chi Sigma; also UAV.275.5, 1911–1912. Baxter wrote numerous letters to other universities seeking comments on the fraternity. W. A. Noyes at Illinois recommended Phi Lambda Upsilon.

Germany and Harvard, 1912-16

Germany and Europe, 1912–13

During his last year of graduate work at Harvard, Adams received a Parker Fellowship for 1912–13. This fellowship carried a stipend of $750, which was adequate to support a year of study and travel abroad.[1] C. L. Jackson, his Harvard mentor, advised him to work with the brilliant Richard Willstätter, if possible, and suggested a semester at Berlin in Emil Fischer's laboratory as another possibility.

Adams went to Europe in June 1912 and traveled in several countries, including Sweden, Finland, and Russia.[2] His love of novelty and adventure shows in the stub of a ticket for a trip from Potsdam to Berlin by zeppelin in Germany;[3] this was undoubtedly his first venture into air travel.

Adams registered for the first semester at the University of Berlin; his study record shows that he attended lectures by Fischer, Diels, F. Sachs, and others. He later described his position as "a postdoctorate student with Otto Diels, then privatdozent in Emil Fischer's laboratory at the University of Berlin, on an unsuccessful attempt to prepare a carbon nitrogen compound, namely dicyanocarbodiimide".[4] He made friends with German graduate students and corresponded for a time with them. It is doubtful if Adams had much contact with Emil Fischer, who was then 60 years old and as busy as Adams was to be in later years at Illinois. Fischer was the undisputed world leader of organic chemistry: as Willstätter described him, a princely man, who towered over all his collaborators in greatness, imagination, and instinct for research.[5]

The rising star in German organic chemistry, Richard Willstätter was in Jackson's opinion the only German organic chemist comparable to Fischer. Jackson said, with the characteristic outlook of age, that when he was a student in Germany in the 1870s, there were at least half a dozen great figures, but now there were only two.

The difficulty in working with Willstätter was that in 1912 he was moving from Zurich to the new Kaiser Wilhelm Institute in Dahlem, a suburb of Berlin, and his laboratory was not ready for occupancy.[6] Adams wrote him several times and persisted in his request to join his group, in spite of Willstätter's attempts to discourage him. An illness that immobilized Adams for several weeks at the beginning of 1913 complicated the situation. Willstätter finally gave in to Adams's insistence, allowing him to join the new laboratory early in 1913 and spend several months there.[7] In a letter to W. A. Noyes in 1916,[8] Adams wrote that he worked with Willstätter on the synthesis of some dipyrrylmethanes related to the structure of chlorophyll, one of Willstätter's major problems. Although he made some progress, he did not get the final products and there are no publications by Adams from Dahlem; apparently the time was too short and perhaps the laboratory not yet efficiently organized. Willstätter's autobiography does not mention Adams, although it names later American workers, including Adams's close friend E. K. Bolton, who was in Dahlem the succeeding year (1913–14). We have found no contemporary documents about Adams's stay in Dahlem, but Adams became a good friend of Willstätter's second-in-command, the Swiss chemist Arthur Stoll, and Willstätter himself wrote Adams a cordial note in 1926 when be became head of the Illinois department.[7]

Adams's other written accounts of his experiences in Germany are from many years later, although he frequently talked about Willstätter's laboratory to his Ph.D. students in the 1920s.[9] His European year strongly influenced his philosophy of graduate education and contributed to his ideas of how to conduct a first-rate department. Adams was one of the least provincial and parochial of men, but he received his entire higher education at Harvard, and experience of foreign countries and of a completely different university system was salutary. Adams's later writings and his work at Illinois showed that he disliked fundamentally one characteristic of the European university system: there was only one professor in each department or institute, who controlled absolutely the activities of junior staff members and research students, as well as the teaching programs. Adams's whole career was based on his philosophy of recognizing and encouraging originality and talent wherever he found it and in helping younger colleagues to develop independent research programs.

Nevertheless, he had found much to admire and use as a model in Willstätter's laboratory. In a lecture in 1966 Adams said:[4]

> My experiences in Germany 1912–13 deserve a few words—first at the University of Berlin and then at the Kaiser Wilhelm Institute in Dahlem. What opened my eyes most was the type of equipment used in the chemistry laboratories. One of the most significant things I learned during my stay in Germany was the importance of having in the laboratory equipment of proper size and shape for the amount of material in hand for a particular reaction, and carefully selected conditions under which to run the reaction. Second most important was use of the chemical library. I remember when Prof. Willstätter asked me to look up something in Beilstein and in Stelzner. I had not heard of either of these

National Academy of Sciences

Charles Loring Jackson (1847–1935).

Illinois University Archives

Richard Willstätter (1872–1942).

National Academy of Sciences

Elmer P. Kohler (1865–1938).

publications. They were not available in the Harvard library—though they had been published two years before I took my doctor's degree. I gained much satisfaction while in Germany in seeing, hearing, and sometimes meeting distinguished organic chemists of those days, pioneers in development of organic chemistry—Von Baeyer, Emil Fischer, Hantzsch, Von Braun, [Fritz] Hofmann, Beckmann, Nernst, Heinrich Wieland, and many others.

While I was with Professor Willstätter he invited twice a semester his 14 post-doctorate assistants to his house for a social evening where he served delicious erdbeeren bohle—a strawberry punch formulated with plenty of potent wine. As the evening advanced and we were all becoming mellow, including the professor, he would start talking about the future of organic chemistry. He was convinced even in those days that the most important advances would be along biochemical lines. And Willstätter himself in the last years of his active career had undertaken research on enzymes.

In pencil notes added to this lecture manuscript, Adams mentions that the desk space available in the laboratory was "meager in terms of today."

To a German correspondent, Adams wrote in 1965[10]:

My experience with Professor Willstätter in 1913 was a most stimulating one. I acquired from him information on attacking research problems and on techniques which were invaluable to me later on. I felt that my sojourn in Germany at that time provided an experience which benefited me during the rest of my career.

Willstätter was a pioneer in the use of platinum and palladium for catalytic hydrogenation of organic compounds. In part of his Ph.D. research, Adams had synthesized *n*-nonane, the last stage being a reduction of nonene with hydrogen over a nickel catalyst at 160°.[11] The Dahlem laboratory increased Adams's interest in catalytic reduction, and later he was to make a most significant contribution to it. E. J. Corey's statement, "In Europe he learned what it took to do first rate research and became familiar, particularly in Willstätter's group, with the style in which large and formidable projects were undertaken",[12] summarizes valuable scientific results of his European year.

Adams applied for a renewal of his fellowship, but it was awarded instead to his friend E. K. Bolton. Apparently the Harvard faculty thought that two men each abroad for one year would learn more than one man in two, an altogether reasonable view. Nevertheless, it meant that Adams needed a job in 1913 when he returned from Europe.

Jackson offered a temporary solution; he wrote Adams on April 24, 1913:[1]

I was sorry to hear that you had not got a fellowship again, but it may be to my advantage, if you will accept a place as my private assistant next year. I shall be glad to pay you eight hundred dollars ($800). This will be clear, as I shall pay

> all the expenses of the research, and of course you could not pay term-bills, even if you wished to.
>
> I hope you will see your way to taking this place. It will give you a chance to wait for a really good permanent place, and we shall have a good time working together, unless the contrast to Willstätter is too trying, which I confess I fear.
>
> Even if I have another assistant (which is very doubtful) you would have first choice of subjects. My ideas are not in order yet, but I think it will be work on the theory of replacement of I or Br by H through sodium malonic ester. [Frank C.] Whitmore has made some progress on this, but I feel sure cannot finish it.

C. Loring Jackson (A. B. 1867, Harvard) studied with Bunsen in Heidelberg and with A. W. Hofmann in Berlin during 1873 and 1874 without taking an advanced degree. His research program at Harvard dealt with aromatic and quinone chemistry, pioneering work in determining the structures of colored products from bases and polynitro aromatic compounds. He also showed that nucleophiles like sodiomalonic ester displace halogen from polynitro aromatic halides, the general problem offered to Adams.

Jackson was giving up active teaching after forty years at Harvard because of poor health, although he lived until 1935. He was a wealthy bachelor, and he undoubtedly planned to pay Adams's stipend and expenses personally. He followed Adams's later career with great interest. Adams accepted Jackson's offer to return to Harvard, and, before he sailed from Europe, an unexpected opening gave him an additional opportunity.

Literature Cited

1. RAA, 4, 1912; this contains most of the contemporary documents about Adams's European year, including C. L. Jackson's letter; also Ref. 7.
2. N. J. Leonard, *J. Am. Chem. Soc., 91,* a–d (1969); this account of Adams's career, written for his eightieth birthday, was checked by Adams personally for accuracy.
3. RAA, 4, 1913–1914; contains the ticket, dated August 28, 1913, for a trip by "Zeppelin-Luftschiff Hansa," of the "Hamburg-Amerika Linie, Abteilung Luftschiffahrt." Count Ferdinand Von Zeppelin (1838–1917) was the pioneer in the development of rigid lighter-than-air craft (zeppelins or dirigibles) and manufactured many after 1900. They were used to bomb London in World War I. The date shows that Adams could not have returned to the United States before September 1913.
4. RAA, 44, Noller Symposium; lecture at the retirement dinner for his Ph.D. student Carl Noller at Stanford, May 20, 1966.
5. R. Willstätter, *Aus Meinem Leben,* Verlag Chemie, Weinheim, 1949, pp. 211 ff. The characterization of Fischer is detailed and masterly.
6. Willstätter's move to Berlin is described in detail in *Aus Meinem Leben,* pp. 198 ff.
7. RAA, 7, Willstätter, 1912–1913, 1926–1927.
8. Illinois Archives, 2/5/15, 364, Roger Adams; RA to W. A. Noyes, August 7, 1916.

9. Account of W. H. Lycan, Adams Ph.D. 1929, in E. J. Corey, *Welch Foundation Conference,* 1976, p. 218.
10. RAA, 34, European Correspondence B; RA to D. Bauer, August 2, 1965.
11. Latham Clarke and RA, *J. Am. Chem. Soc., 37,* 2536 (1915); the method of preparing the nickel catalyst was not given, but it was presumably Sabatier's procedure, P. Sabatier and J. B. Senderens, *Compt. Rend., 124,* 1358 (1897) and many later papers.
12. E. J. Corey, op. cit., p. 208.

On the Harvard Faculty, 1913–16

The division of chemistry at Harvard to which Roger was returning was passing through critical times.[1] Torrey's death in 1910 deprived organic chemistry of a promising teacher, and in September 1911 George L. Kelley, who had received his Ph.D. with Jackson in June and had two years of experience as an Austin Teaching Fellow, was appointed instructor.[2] He gave all the organic courses offered: Chemistry 2, the elementary half course (no laboratory); Chemistry 5, the carbon compounds (advanced lectures and laboratory); and the two advanced half courses, Chemistry 16 and 17 in organic reactions and special topics.[3]

In 1912 the division suffered a double blow in the untimely death of Charles R. Sanger (Ph.D. 1884, Harvard), professor of qualitative analysis and director of the laboratory, and the withdrawal of C. L. Jackson from his active role in teaching Chemistry 1 and directing research in organic chemistry. Thus G. P. Baxter, chairman of the division, after consulting with the remaining permanent members, T. W. Richards and L. J. Henderson, wrote to President Lowell that it would be necessary to spend $9,200 to $11,000 out of his proposed department budget of $31,200 to hire four new men.[4]

In the spring of 1912 Baxter was able to hire Elmer P. Kohler (Ph.D. 1892, Johns Hopkins), chairman of the chemistry department at Bryn Mawr, a productive research worker and highly recommended by Remsen as an unusually good lecturer, to teach the beginning inorganic course and Chemistry 5. Grinnell Jones (Ph.D. 1908, Harvard) was brought back from the University of Illinois where he was an instructor, and Arthur B. Lamb (Ph.D. 1904, Harvard) returned to Harvard from his position as head of the chemistry department at New York University. Arthur Michael, a leading organic chemist in the United States and retired from Tufts, also joined the Harvard faculty. Kelley, although he had been considered a successful teacher during the year, was not to be given a permanent position. Baxter planned to replace him "only on grounds of inexperience and lack of reputation" but renewed his appointment for a second year.[4]

During the summer Kelley accepted a position with Midvale Steel and Ordnance Company.[5] Although he informed Baxter and Richards of his move, his resignation, effective September 23, 1912, allowed little time for replacement. Fortunately, James Norris (Ph.D. 1895, Johns Hopkins), one of Remsen's students, was teaching at nearby Simmons College and agreed to accept a part-time lectureship at Harvard[5] to give Chemistry 2 and Chemistry 17.

This temporary arrangement worked out well; Jackson commented that Norris had "made a gigantic success of the course last year,"[4] and it was continued for 1913–14. Chemistry 16 was still open, and in a letter dated July 17, 1913, and sent off to Europe to try to catch Adams abroad, Baxter offered Roger an appointment:[6]

> As you are to be in Cambridge next year, it has occurred to us that you might like to give Chemistry 16' hf. on the progress of organic reactions. You would

have the title of Instructor and there would be a small salary which cannot be decided till fall. Professor Norris is to give 2 & 17 again. If you will do this, it will be a great service to us and should help you in obtaining a position for the following year. Will you let me know as soon as possible how you feel about it. Professor Jackson approves.

I can let you have Dr. Kelley's lecture notes in 16.

Baxter's offer delighted Jackson. In a letter dated June 23, 1913, from Pride's Crossing, he wrote Baxter:[4]

I think your plan is a first rate one for Adams and me—for A., as it will help him to get a good place, for me as it will anchor A. with me for the year. I confess I do not understand how a man can give 16, who had not had years of experience in research, but I suppose he can work up something decent from the textbooks.

"A." accepted the two positions and returned to a situation that was a far cry from Dahlem and other leading laboratories in Europe. Boylston Hall, for many years the only chemistry building at Harvard, had been built in 1858 for only forty laboratory students. It was repeatedly remodeled and enlarged by "one dreary makeshift or another" because of growing student enrollment in chemistry, which topped 600 by 1912. The struggle to provide laboratories for the 400 students in Chemistry 1 and for the students and faculty in exacting physical and analytical work left little space for organic chemistry, and no elementary laboratory instruction was possible until the senior year in Chemistry 5. An organic research room for graduate students and the private laboratories of faculty members were available for graduate research. Arthur Michael maintained his own private laboratory in his home in Newton Center. Library facilities were also poor, and many standard works that Adams had used in Germany, including *Beilstein*, were not at hand. In 1913 the opening of the Wolcott Gibbs Memorial Laboratory for Richards and his group and the T. Jefferson Coolidge Laboratory for analytical chemistry offered some relief.

Adams joined a department staffed amost entirely by Harvard men in the traditional fashion. T. W. Richards (Ph.D. 1888, Harvard) and his pupil, G. P. Baxter (Ph.D. 1899, Harvard), were performing the work on atomic weight determinations that attracted worldwide attention and experienced graduate students to work in this field. Three other faculty members were Richards men, G. S. Forbes (Ph.D. 1902), and Jones and Lamb, all in physical chemistry. L. J. Henderson (A.B. 1898, M.D. 1902, Harvard) taught biochemistry. For years the organic chemists had been Harvard graduates: H. B. Hill, Jackson, and Torrey. The appointments of Kohler and Michael were almost the only permanent outside appointments made in chemistry in Harvard College since its founding.

In his first year Adams did research for Jackson and gave Chemistry 16; he was reappointed for 1914–15 and taught all the organic courses offered except for Kohler's Chemistry 5. It was at this time, with the long-awaited laboratory space available, that laboratory instruction to accompany the elementary lecture course became possible, and Adams organized and taught the new one-semester Chemistry 22, which was limited to twenty-five students. He also completed the research with Jackson on hexabromodiacetyl and published this and his earlier work with Clarke on *n*-nonane in early 1915.[7]

Using the new laboratory space, Kohler initiated senior research, and Adams directed Henry Gilman (Ph.D. 1918, Harvard) in the synthesis of substituted phenyl esters of oxalic acid, demonstrating the new and useful reagent oxalyl chloride in well-planned and well-executed experiments. They completed and published the research within the year,[8] and Gilman later wrote of this experience:[9]

> In my senior year at Harvard, I was interested in doing some orienting research with him [Adams]. This was done just as an aside, and either carried no credit or only a small token of credit. Not a little of the work was done at night, and I recall how when the research was completed for the day (often near midnight) we would cross the street to a drug store on Massachusetts Avenue for a chocolate malted milk.
>
> The experiments were a great delight for me, and he would come in somewhat frequently for chats. He was of course most friendly, interested and helpful. The study was not "monumental," but it was exciting for each of us: his first direction of research, and my initiation into research.

Gilman added that this small publication of directed research was a significant factor in Roger's appointment at Illinois.

Roger's sixth class reunion came in June 1915 and he reported in his class book, as he was careful to do all the years.[10] By then 151 of his classmates had received advanced degrees, 71% of these from Harvard, including eleven Ph.D.'s from Harvard with two in chemistry, Roger and G. Esselen, who was with Chemical Products Company in Boston. Twelve men indicated their profession as chemistry, a calling that was distinguished by its low pay; with five chemists reporting, the average salary was $1,500, the second lowest (ranking above agriculture) out of the twenty-three occupations tallied. The class report noted changes in Harvard under President Lowell's administration and from six years' distance: scholarship was entirely respectable and A's were admirable; chemistry labs were better; men drank less, worked harder, and dressed worse; authorities and requirements were stricter; the elms were gone and the Yard was "barren"; freshman dorms and the Lars Anderson bridge to the stadium had been built; the subway to Boston was good—"almost too easy to get to town and out at any hour."

During Roger's late summer vacation, after teaching summer school, he toured the West, meeting a friend in Spokane, Washington, and camping in the Canadian Rockies to hunt and fish. He returned through California and the Southwest visiting San Francisco, San Diego, and the Grand Canyon. A family photograph shows the young adventurer riding on muleback down the Bright Angel Trail (Grand Canyon) on September 20, 1915.[11]

Back at Harvard the next year, he followed the same teaching schedule as before. Kohler (elected the Abbott and James Lawrence Professor of Chemistry in 1914) and Adams had both become very popular teachers of the elementary inorganic and organic courses, respectively, and the number of students was rising. The elementary organic laboratory course that Adams and two assistants gave overflowed, and they repeated it in the second semester. Adams boldly wrote Baxter detailing the organization and work load of Chemistry 2 and 22 and their inconvenient quarters and calling for at least two more "absolutely necessary" assistants for safety and proper instruction.[12]

James B. Conant later recalled Adams's success as a teacher:[13]

> When he took over on short notice the elementary course in organic chemistry at Harvard, he immediately proceeded to enliven it by anecdotes and by many more lecture table experiments than in the past. He was a hit from the outset and I used to hear a good deal about his brilliant and colorful presentation of the difficult subject of organic chemistry. I remember particularly explosions with acetylene and oxygen, which were then something of a novelty for table demonstrations. Since I had to succeed him in that class, again on short notice, I am painfully aware of what a reputation he left behind him when he moved from Harvard to Illinois.

In addition, Roger directed two research problems with master's candidates, continuing the study of the action of oxalyl chloride on alcohols with L. F. Weeks, and examining the reaction of the Grignard reagent on nitriles to synthesize amidines with C. H. Beebe. They completed the work that summer and promptly wrote it up for publication.[14] During these years his friendship with Kohler and with Jackson grew; these two followed his career with great interest.

In January 1916 he accepted his fourth appointment as instructor for the academic year 1916–17, but late in the summer a position at the University of Illinois became available. In a long hand-written letter to W. A. Noyes, head of the chemistry department, Adams described his activities at Harvard and his outlook.[15] At Harvard he was teaching two elementary and two advanced courses besides doing research, and at Radcliffe he gave "an elementary course in organic chemistry and every other year an intermediate course, each of these Radcliffe courses having two lectures a week." For two weeks in June he aided in laboratory entrance exams for Harvard and graded papers for the College Entrance Examination Board. He added several hundred dollars to his stated salary by teaching summer school.

His current research, in addition to the problems already published, included work on the reactions of Grignard reagents with alkylthiocyanates, acyl cyanides, and related compounds; the synthesis of *o*-amino azo dyes (related to his Ph.D. thesis); and the investigation of the relative reactivity of an atom on a carbon alpha to a negative group. The study of ring closure between the meta and para positions in benzene derivatives, although supposedly ready for publication "next fall," was actually published from Illinois in 1923,[16] the forerunner of his extended work on diphenyls. Under his direction a student was synthesizing aliphatic 1,5-diketones to prepare the substance obtained by C. Harries in Germany by ozonizing natural rubber. "Of course, I have many other subjects I'd like to try out, but I'm more interested in these at present and they will serve to show you the general type of organic chemistry which appeals to me most."

Up to now his research workers had been inexperienced, but for the next year he had already signed up four good students to do research more than half time, and he expected one or two more, with a growing research group in the coming years.

From this account of his teaching and research activities, one can see the extraordinary energy and ability with which Adams had seized his opportunity at Harvard, in spite of a very heavy teaching load. His subsequent accomplishments at Illinois seem more credible when viewed in the light of his three years at Harvard.

Literature Cited

1. S. E. Morison, ed., *The Development of Harvard University,* Harvard University Press, Cambridge, 1930. Chapter XVI, "Chemistry," by C. L. Jackson and G. P. Baxter, pp. 258–76, supplies information except as noted specifically.
2. Harvard Archives, UAV. 275.5, Dept. of Chemistry Correspondence, July 1910–11; Pusey Library, Harvard University. Henceforth references to these archives will be given by the number above and date of the box of correspondence.
3. All references to course offerings and their instructors are from the appropriate year of the Harvard University Catalogues, from 1911–12 through 1916–17.
4. UAV. 275.5, June 1911 to September 1912; G. P. Baxter to President Lowell, March 11, 1912, and additional correspondence.
5. *American Men of Science, 1921.*
6. RAA, 64, 1913–64.
7. RA and C. L. Jackson, *J. Am. Chem. Soc.*, *37*, 2522 (1915); RA and L. Clarke, ibid., 2536.
8. RA and H. Gilman, ibid., 2716.
9. H. Gilman to DST, May 9, 1977.
10. *Sexennial Report, Class of 1909, Harvard College,* Sexennial Report Committee, Cambridge, 1915.
11. RAA, 4, 1915–16; RA to C. L. Jackson, August 4, 1915. L. A. Brink papers, Photo Album.
12. UAV. 275.5, October 1915 to April 1916: "Report of Overseers written for Richardson" in Baxter's handwriting; RA to Baxter, Jan. 19, 1916.
13. RAA, 21, Conant; J. B. Conant to R. M. Joyce, July 13, 1954.
14. RA and L. F. Weeks, *J. Am. Chem. Soc.*, *38*, 2514 (1916); RA and C. H. Beebe, ibid., 2768.
15. Illinois Archives 2/5/15, 364, Roger Adams; RA to W. A. Noyes, August 7, 1916. Baxter had written that he was glad to have Adams stay but he was "under no obligation." Ref. 12.
16. RA and W. C. Wilson, *J. Am. Chem. Soc.*, *45*, 528 (1923).

Move to Illinois, 1916

The appointment of Roger Adams to the chemistry department at Illinois in 1916 was one of the most important ever made by the university; few appointments made by any American university have had more beneficent results for American science.

The University of Illinois was founded in 1868 and occupied "a most uninviting strip of flat open prairie" 1 mile from Champaign and Urbana.[1] The first professor of chemistry to function was A. P. S. Stuart, trained by Josiah Parsons Cooke at Harvard. Stuart started laboratory instruction and collected a well-chosen chemical library, but he resigned in 1874 apparently because of lack of funds and support. A chemical laboratory was built in 1878 costing $40,000, but in 1882 both teachers of chemistry, one a student of Justus Liebig, left the university. In the ensuing years, A. W. Palmer, an Illinois graduate, taught at Illinois for a time, took a doctorate at Harvard, returned to teach at Illinois, and then went abroad to study further at Göttingen with Victor Meyer and with Hofmann at Berlin. Recalled to Illinois to take over as assistant professor after a collapse of instruction in chemistry, Palmer worked hard with two assistants to maintain a strong program. In 1890 S. W. Parr was appointed, and in 1894 he was designated professor of applied chemistry. This was the origin of the unusual arrangement by which industrial chemistry, later titled chemical engineering, became a part of the chemistry department, an arrangement that continued. Parr developed the combustion calorimeter and later the Parr shaker, used particularly for catalytic hydrogenation with Adams platinum catalyst. Palmer also assumed the direction of the State Water Survey, established in 1895.

A fire started by lightning badly damaged the laboratory in 1896, and Palmer and his hard-pressed colleagues did not get a new building until 1901, when the west section of what is now the W. A. Noyes Laboratory was built. Palmer died suddenly in 1904, apparently from overwork, although some research he managed to do on arsines may have aggravated his condition. He had created a solid foundation for further teaching and research in chemistry,[2] and H. S. Grindley, a nutritionist with an Sc.D. from Harvard, a member of the department since 1895, carried on as director of the laboratory.

In 1904 Edmund J. James (1855–1925) became president of the university; he had obtained a Ph.D. in economics from Halle in Germany, had been a leader in the Wharton School at Pennsylvania, and was president of Northwestern before 1904. He was an energetic and strong-willed man, ambitious to raise the standards of faculty and student scholarship and to make Illinois a recognized center of learning. He was an accomplished politican in dealing with the faculty and with the Illinois State Assembly; indeed, he fancied that he had a chance for the Republican nomination for the presidency in 1916, apparently encouraged by the political success of another university president, Woodrow Wilson.[3]

National Academy of Sciences *William A. Noyes (1857–1941).*

Illinois University Archives

The completed Illinois chemistry building (1916).

James's drive for excellence and his conviction that good research and teaching were not only compatible but necessary to each other made him seek William A. Noyes, then at the Bureau of Standards, as director of the laboratory. Eventually James was successful, and Noyes was inaugurated in 1907 in ceremonies reminiscent of the German universities, which James admired passionately.

Noyes brought to Illinois a high reputation as a research worker, teacher, writer of textbooks, editor, and public-spirited citizen of both the scientific and the human community. His mild appearance covered a rocklike will and very high standards of performance for himself and others. Noyes was a good judge of people and an indefatigable worker, with almost no sense of humor. He was highly respected and his recommendations carried weight with deans, presidents, and trustees. Noyes attracted graduate students to Illinois and built up an excellent staff in the various branches of chemistry.

Formal ceremonies in 1908 opened the graduate school at Illinois as a separate unit; it received an annual sum of $50,000 per year for two years, said to be the first appropriation in a state university made specifically for graduate work.[3] In 1907 there were eleven senior staff members in the chemistry department and seventeen graduate students. In 1926, when Noyes retired, the senior staff numbered twenty-five with about 120 graduate students. The number of departmental scientific publications increased in even greater proportion; thirty-five publications, including two books, appeared in 1915.

In 1915–16 departures of key faculty in several fields of chemistry severely threatened the department. Later correspondence of Noyes and Adams with James suggests that these departures were probably due to low salaries and slow advancement in chemistry in spite of Noyes's best efforts. The senior organic chemist next to Noyes, C. G. Derick, resigned to take an industrial research position with the growing Schoelkopf Aniline and Chemical Company. An Illinois Ph.D. with Noyes (1910), he published interesting synthetic work and also papers in physical organic chemistry on strengths of organic acids, which attracted much attention. He started the organic chemical manufacture ("preps," to be discussed later) in the summer of 1914. Derick's departure reduced the organic contingent to Noyes and an instructor, Oliver Kamm.

C. W. Balke, the leading inorganic chemist at Illinois, 1907–16, with a Ph.D. from Pennsylvania under Edgar F. Smith, accepted an industrial position with the Pfannstiehl (later Fansteel) Company of North Chicago, which specialized in alloys and products derived from some of the less common elements such as columbium and tantalum. E. W. Washburn, a Ph.D. from MIT and the senior physical chemist from 1908 to 1916, gave unmistakable signs of dissatisfaction. Although he stayed at Illinois until 1922, it was in ceramic chemistry and ceramic engineering, and his replacement in physical chemistry was needed.[4]

Noyes reacted to the disintegration of his senior staff with his customary energy and tenacity. The erosion of faculty was partially offset by a favorable development for Illinois chemistry in 1916, the completion and dedication in

April of an addition to the 1901 building. This gave a total of over 160,000 square feet of floor space for chemistry and housed all chemical research (including biochemistry and chemical engineering) during almost all of Adams's time at Illinois. (The chemistry annex, completed in 1931, housed mainly freshman teaching.) The first section of East Chemistry (later the Roger Adams Laboratory) was dedicated in 1951.[5]

The dedication of the new section came during a national spring meeting of the American Chemical Society, which Noyes had arranged for Champaign–Urbana with elaborate plans.[6] A torrential downpour enlivened, or at least signalized the actual dedication of the new structure, washed away the interminable oratory so highly regarded at that time, and sent all the dignitaries and spectators scrambling for cover.

A press release, distributed to the newspapers of the state, was undoubtedly written by Noyes at the request of James. It described the content, growth, and significance of the program in chemistry, as well as the new building, and showed Noyes's clear view of the need for trained chemists in the chemical industry:[7]

> In still greater degree it [the growth of chemistry] is connected with the very rapid development of Industrial Chemistry and of Chemical Engineering in America during the past decade. Scores of establishments which would never have thought of employing a chemist fifteen or twenty years ago now find that men thoroughly trained in this line of work are indispensable for their success. . . .
>
> The war in Europe has called the attention of the public to our dependence on Germany for many kinds of chemical products. This will undoubtedly prove a great stimulus to many lines of manufacture in which we are now deficient and this, in turn, will create an increased demand for chemists. The increased facilities which the addition will afford will make it possible for the University of Illinois to do its full share in supplying this demand.

It is amazing that the 1901 section of the building, built entirely of wood except for a brick facing, did not burn down in one of the perennial laboratory fires. When the old laboratory was rebuilt inside after World War II, construction workers found, as Bailar states,[2] that it had heavy brick lateral fire walls and the floors had heavy beams with a twelve inch layer of sand between the ceiling of one story and the wooden floor of the next. These precautions may have helped preserve the 1901 building.

To rebuild his department, Noyes offered a professorship in physical chemistry to Joel H. Hildebrand of Berkeley, who politely but promptly refused. He considered Elliott Q. Adams of Berkeley, one of G. N. Lewis's students, who had published outstanding work on dyes. However, he hired Richard Chace Tolman of Berkeley, who came to Illinois at $3,000 per year but stayed only two years, moving to Washington after the nation entered World War I. Both Hildebrand and Tolman had distinguished subsequent careers at Berkeley and Caltech, respectively. Horace G. Deming (Ph.D. 1911, Wisconsin) accepted the appointment in inorganic chemistry at the rank of associate; he stayed in Urbana two years and then moved to the University of Nebraska. (Illinois was one of a small number of universities that, until World War II, had the rank of "associate" between the level of instructor and assistant professor.)

The organic appointment was more troublesome. After considering Moses Gomberg from Ann Arbor and L. H. Cone, Gomberg's associate in his free-radical work, Noyes made an offer to J. M. Nelson (Ph.D. 1907, Columbia, with M. T. Bogert, and a faculty member there). Nelson had published a series of papers on purification and study of enzymes; what particularly interested Noyes in him was undoubtedly a group of papers on the electronic structure of molecules, published around 1910 with K. George Falk of MIT, later of the Harriman Research Laboratory of the Roosevelt Hospital in New York. These papers utilized the "dualistic" theory of valence, using plus and minus signs for each bond, which was advocated particularly by H. S. Fry of Cincinnati, Julius Stieglitz of Chicago, and Noyes. It was generally replaced by G. N. Lewis's electron pair theory of valence of 1916–23, but it persisted into the 1920s. Nelson refused the Illinois offer on August 16, 1916, and stayed at Columbia for the rest of a long career. He did excellent research on oxidizing enzymes and was beloved by students and colleagues alike, being known as "Pop Nelson."

Noyes was now (mid August 1916) at his summer home in Frankfort, Michigan, and he bombarded Dean Babcock with sheaves of nearly illegible long-hand memoranda about what to do next. Adams was already a prospect, since Noyes had written him on August 1, informing him of the position and inviting him to return an application form. Adams accompanied his application with the detailed letter to W. A. Noyes of August 7, 1916, quoted earlier, enlarging upon his experience and prospects at Harvard and stating his interest in the position "provided it is an assistant professorship." Adams's prospects for good graduate students and promotion at Harvard were excellent, and he was therefore speaking to Noyes from a strong position. The rapid emergence of Conant (who was essentially Adams's replacement at Harvard) as a research chemist shows that Adams was not exaggerating to Noyes his own future in Cambridge. Unfortunately, the archives of the two universities do not contain the letters of recommendation for him, but he was unquestionably warmly supported by Kohler, Jackson, and probably Richards. Noyes turned immediately to Adams after Nelson's refusal.

The following telegrams and letters indicate the negotiations with Adams:[8]

Telegram—(Dean) K. C. Babcock to W. A. Noyes—August 17, 1916:

President authorizes appointment of Roger Adams as assistant Professor at twenty-seven fifty. Am wiring him....

Telegram—K. C. Babcock to Roger Adams—August 17, 1916:

University of Illinois offers you assistant professorship of chemistry for three years beginning September 1 at initial salary of twenty-seven hundred and fifty dollars. Summer session pay for assistant professor three hundred. Wire reply collect at earliest convenience.

Telegram—Roger Adams to K. C. Babcock—August 23, 1916:

Telegraphed Noyes Monday following questions what is normal number lectures weekly would I have choice of all organic courses and afterward free hand in conducting them what is usual rate of promotion for successful members of staff no reply received must hear before deciding need vacation don't wish to leave till decision made please wire as soon as possible.

Telegram—K. C. Babcock to Roger Adams—August 24, 1916:

Noyes assures choice of organic courses. Free hand in conducting them. Four to seven hours of lectures. Quizzes and conferences. No scale of promotions. Success in teaching and research promptly recognized. Washburn made full professor after three years as assistant professor.

Telegram—Roger Adams to K. C. Babcock—August 24, 1916:

Your telegram received. Will accept with pleasure Illinois position under condition named, and in addition what I could expect at Harvard namely a yearly advance in salary of one hundred dollars and if reappointed an advance in grade.

Telegram—Roger Adams to E. J. James—August 27, 1916:

Will agree to offer of twenty-eight hundred instead of my plan. Am still troubled about promotion. Have learned from reliable sources rapid promotion at Illinois chiefly obtained by receiving calls elsewhere. Unwilling to submit to this plan. Must receive absolute assurance from you my success in teaching and research will be recognized by promotion independent of more or less accidental outside offers. Wish to devote entire energy to teaching and research at Illinois in full confidence success will bring its just reward. My proposal of discharge or promotion to higher rank in three years based on this. Leaving Harvard means rejecting practically certain assistant professorship in three years at much higher salary than Illinois offers now.

At this point President James entered the scene, clearly feeling that this brash young Harvard man needed firm handling. Noyes had meanwhile written Adams that he would have essentially a free hand in teaching and research; the text of the letter can only be inferred from other extant documents. James wired Adams on August 28 and wrote him the next day.

Telegram—E. J. James to Roger Adams—August 28, 1916:

Have no authority to give you such assurances as your telegram calls for. You must take your chances with the rest of us. Still hope you will come. Letter follows. Remember me to Dr. Dewey.

Letter—E. J. James to Roger Adams—August 29, 1916:

We cannot undertake to alter our entire system of University management in order to meet the wishes of any particular man. We should be very glad to have you accept the position which we have offered you at the salary indicated and for the term indicated. I have no doubt that if you succeed in your work you would be promoted here as rapidly and with as much increase in salary as you would be promoted elsewhere. But that is, of course, a prophecy and not a promise. I make it a rule never to make promises as to what the University will do in the future. Any one of us may drop out of work long before the times comes to fulfill such promises and the real security, after all, lies in one's ability, in one's success, and in the general situation.

In all human probability the University of Illinois will continue to grow and prosper and yet it is perfectly possible that the legislature of Illinois, by refusing appropriations, should put an end to its existance. As we have no certainty or promise from the legislature that appropriations will be continued, everyone takes his life in his hands to that extent.

I was greatly pleased at the statements I heard about you and your work, and we should be greatly pleased to have you with us.

Adams's prospects at Harvard and the poor record of Illinois in losing three of its senior chemists within a year justified his insistent question about promotion. Adams's request was clearly advised by his good friend E. P. Kohler, a shrewd Pennsylvania native, who in 1911 had forced T. W. Richards and President Lowell of Harvard to make him an acceptable offer to move from Bryn Mawr to Harvard after an initial proposal that was almost insulting.

Adams was not passive during these negotiations. His friend James F. Norris, who had taught as lecturer in chemistry at Harvard in 1912 and 1913, wired him on August 21, 1916, from his vacation place in Maine:

Unusual opportunities research Illinois fine laboratory everything required furnished many students good spirit in department important question living Boston or small town broaden experience by going more ahead advise Illinois congratulations.

Kohler, in his letter of August 30 from Lake Louise, Canada, gave him the same advice, with obvious personal regret:[9]

I never did anything more reluctantly than send the advice I did; but I thought over the matter all through a long walk and never once could see it any other way. As there is no doubt that the president agreed to grant your requests about promotion, I have no doubt that you have accepted so hereby wish you all possible luck in your new field, confident that you will succeed and that sooner or later we shall have you back.

Too bad that you decided not to come [to Lake Louise]. I had picked out a beautiful place for our first camp and looked forward to a few weeks of real mountain work with a great deal of pleasure.

C. L. Jackson's later advice agreed with Kohler's.

Although President James did not yield on the promotion question, as Kohler thought he would, Adams accepted on September 1. James was still feeling somewhat irritated, as his subsequent letters to Adams and Noyes show. Neither James nor Noyes could know that this appointment was one in a thousand, although Kohler, and Adams himself, may have had premonitions of his great accomplishments at Illinois.

Telegram—Roger Adams to K. C. Babcock—September 1, 1916:
Accept Illinois position. Please write as soon as possible to Boylston Hall, Cambridge the date of opening of Illinois and date when lectures start.

Letter—E. J. James to Roger Adams—September 2, 1916:
I received your telegram announcing that you accepted the position of assistant professor of chemistry at the University of Illinois. The Secretary of the Board of Trustees will send you soon a notice of your appointment.
In answer to your telegram, I beg to say that the entrance examinations at the University of Illinois are held from September 11 to September 15. The registration days are September 18 and 19—Monday and Tuesday, I believe. I think there is always a general faculty meeting of the College of Liberal Arts and Sciences on the Friday before the Monday of registration, at which all the members of the faculty are expected to be present. That would make it September 15.
I would suggest that you get on the ground as soon as convenient, so that you can find a place to live—which is not always easy here on the prairies—and get yourself comfortably settled before the opening of the year.

James wrote Noyes on September 2:

I have just received a telegram from Roger Adams, accepting the position of assistant professor of chemistry in the University of Illinois for the term of three years from September 1, 1916, at a salary of twenty-eight hundred dollars ($2,800.00) per annum.
I had considerable correspondence with Mr. Adams. He was quite insistent that I should give him assurance of promotion, and all that sort of thing, which I declined to do. I am sending you copies of our correspondence, so that you may have it before you in case any difficulty should ever arise. I told him that I had no authority to give him such assurances as one of his telegrams called for, and that if he came into the staff, he would have to take his chances along with the rest of us.

In his reply of September 13, 1916, to another letter from James, Noyes pointed out that Adams had some basis for his insistence:[8]

I am in most hearty accord with your letter of September 11th about Dr. Adams. I think it is perhaps fair to Dr. Adams to say that his correspondence with Dean Babcock and yourself, which dealt primarily with the business question involved, gives a decidedly one sided view of his attitude toward the question of coming here. In his telegram to me he emphasizes primarily questions as to the character of the work and the degree of independence which he would have in developing our organic division here. In my reply I went into a good deal of detail with regard to the number of graduate students in the division, the amount of lecture and other work which he would be expected to carry, and I assured him of very complete independence in the development of the division along lines that seem to him wise. It is clear from his letter of August 29th, which I enclose, and which I think you have not seen, that my letter was a large factor in the conclusion which has been reached.

There is rather more truth than I could wish in the statement of one of Dr. Adams's telegrams that promotion here has been partly dependent upon securing desirable offers elsewhere. It seems impossible to avoid such a situation entirely, but it is always my desire to promote the men of this department in proportion to their deserts rather than in proportion to what outside parties think that they are worth. It is not surprising that a man who is coming here should feel, in the light of our experience of the past summer, that the salaries which we pay lag a good ways behind the value of the men.

Adams's letter of August 29 to Noyes (the following extract) is sufficient commentary on James's statement to Noyes, in a letter of September 11, that "when a fellow as young as Doctor Adams is primarily concerned about advances in position and salary instead of opportunities for work, I do not think nearly as much of him, from a purely scientific point of view." Adams wrote:[8]

Your letter of August 22 has just arrived and I was very glad to receive it. If it had come earlier I would undoubtedly have worded my telegram to President James differently. I am now simply awaiting the letter which President James is now sending, before I come to a final decision. I will then telegraph Dean Babcock immediately.

My vacation here at Lake Placid is certainly reviving me and in a week or so I'll be ready for work again.

The decision to move to Urbana was undoubtedly the most important one Adams ever made, more fateful than his later ones to decline offers, including ones from Harvard and MIT in 1934, to leave Illinois. Harvard as an institution was not only the oldest university in the country but one of the leading ones; it had been thoroughly revitalized by Charles W. Eliot during his long presidency (1869–1909), and the new president, A. Lawrence Lowell, was equally vigorous and forward looking. The Harvard division of chemistry was headed by T. W. Richards, who had the added luster of the first Nobel Prize ever awarded an American. Although he was an eminent chemist, Richards was a very poor judge of men in general and was constitutionally averse to vigorous discussion and controversy.[10] Kohler, the senior organic chemist at Harvard, was a masterly teacher and experimentalist,[11] and in 1916 he was fifty-one years old, eight years younger than Noyes.

Adams had none of the parochialism sometimes shown by Harvard graduates. He had traveled in the western states, had spent a year in Germany, and had traveled widely in Europe. He realized, with the clear dispassionate insight that characterized him, that there were many promising academic positions outside Harvard. Although his teaching at Harvard was spectacularly successful and his published investigations showed promise in 1916, he scarcely possessed an established research reputation at that time. The Illinois offer would allow him to make a fresh start on his own in a department that, under Noyes, was a recognized and productive center of teaching and research. Adams wished to accomplish something notable in science, and he said that his ambition was to do some research fundamental enough to be cited in elementary textbooks.[12] In 1934 Adams phrased his aim in life thus: "My chief effort has been to contribute something worthwhile in organic chemical research, and in addition in recent years to organize the chemical courses so the students would receive the most efficient training."[12]

The new laboratory facilities at Illinois had received wide publicity and contrasted most sharply with those at Harvard for organic chemistry in old Boylston Hall. Adequate laboratories would be unavailable at Harvard until the Mallinckrodt and Converse buildings arose in 1929; the new Illinois laboratory was undoubtedly an attraction. Adams and Kohler certainly knew of it and of the $5,000,000 appropriation by the Illinois legislature for the next biennium in 1915, said to be the largest grant made by a single law to any university in the country. This support indicated the potential of the university and the chemistry department.[7,13]

Adams probably foresaw to some extent the dramatic expansion of academic and industrial chemistry that would occur in the 1920s. He was undoubtedly aware of the strong chemistry departments at Chicago, Michigan, and other midwestern universities, and the emerging prominence of G. N. Lewis's group at Berkeley. He was never given to looking back at the past, and he approached the Illinois scene with the fixed idea of making an outstanding personal career there and of increasing the standing of the chemistry department. Always a realist, Adams realized that there was an inevitable connection between his own success and that of the department as a whole. His experiences at Harvard and in Germany gave him very clear ideas of how a strong undergraduate and graduate chemistry department should be run, and from September 1916 he worked unceasingly to implement these ideas.

Literature Cited

1. There is no satisfactory history of the University of Illinois, except for a single volume that carries the story only to 1892. The present account is based mainly on documents

in the University Library Archives in Urbana and, unless otherwise indicated, are in the Roger Adams Archive (RAA). Other series have been used and the archival identification will be given when series other than RAA are quoted; box numbers and folder titles are given as in RAA.

2. University of Illinois Archives 15/5/0/1, 1, Departmental Histories; S. W. Parr, in *University of Illinois, Circular of Information of the Department of Chemistry,* 1916, pp. 16–29, Liberal Arts and Sciences. RAA, 54, Speeches; address by Adams on dedication of East Chemistry Building, March 30, 1951. J. C. Bailar, Jr., *J. Chem. Educ., 24,* 550 (1947).
3. Richard A. Swanson, *Edmund J. James 1855–1925: A Conservative Progressive in American Higher Education,* Ph.D. Thesis, University of Illinois, 1966. Swanson's characterization of James agrees with that resulting from examining many letters from him to faculty members and to his rather pedantic bachelor dean, Kendric C. Babcock. James wrote an account of his presidency: *Sixteen Years at the University of Illinois,* University of Illinois Press, Urbana, 1920; mainly a statistical account of the growth in numbers of students and in appropriations during his administration. The graduate school dedication is described in *Science, 27,* 394 (1908).
4. Noyes's correspondence with Dean Babcock, President James, and with several outside chemists documents these departures: Illinois archives, James General Correspondence, 2/5/3, 95, W. A. Noyes; 75, Babcock, May–August 1916; Dean's Office, Dept. and Subject File, 15/1/1, 5, chemistry. Biographies in *American Men of Science* give specific information. G. D. Beal, "Ten Years of Chemistry at Illinois (1916–1927)," in *Special Circular of the Department of Chemistry,* 1916–1927, pp. 9–13, series 15/5/0/1, 1, adds some details.
5. Details are given by Parr and by Adams, Ref. 2.
6. Much correspondence, such as a letter, March 28, 1916, Noyes to James, James, General Correspondence, series 2/5/3, 95, Noyes, about the receiving line.
7. James, General Correspondence, series 2/5/3, 60, Chemistry Laboratory; the dedication was announced in *Science, 43,* 383 (1916).
8. James, General Correspondence above, 130, Noyes.
9. RAA, 7, Kohler; for Norris: RAA, 4, 1915–1916.
10. J. B. Conant, biography of Richards, *Biog. Mem. Nat. Acad. Scis., 44,* 251 (1974); Conant, p. 28.
11. Conant, pp. 35–7.
12. Conversation with D. S. Tarbell (hereafter DST) (1938). The quotation is from Adams's statement written for *Class of 1909, Twenty-Fifth Anniversary Report, 1909–1934,* Harvard University Press, Cambridge, 1934.
13. *Science, 18,* 31 (1903); *42,* 158 (1915), on state appropriations at Illinois.

Illinois, 1916-26

Life at Illinois, 1916–26

The ten-year period 1916–26 covers the time from Adams's move to Illinois in September 1916 to his becoming head of the chemistry department in 1926 upon the retirement of William A. Noyes. Arriving at Illinois in time for the faculty meeting of September 15, he plunged into his professional year with zest: teaching the laboratory course in organic chemistry, accepting his first Illinois candidate for a Ph.D., amplifying investigations on alkali-insoluble phenols begun at Harvard, and initiating a new program of research on local anesthetics in collaboration with Oliver Kamm. At the end of the academic year Adams took charge of the Illinois laboratories engaged in synthesizing organic reagents. This decade saw his service in World War I and the flowering of his capabilities in teaching and research. Later sections consider these events.

Adams rented lodgings at 1007 West Oregon Street, Urbana, where he lived until he moved to Washington for his war service.[1] This location was only a few hundred yards from the chemistry laboratory and very near the site on which the new Roger Adams Laboratory of Chemistry would stand some fifty years later. From the amount of work he accomplished at Illinois, he probably spent little time in his new quarters. He had always lived with his parents and sisters, except for the year he spent in Europe, but his father's death in December 1916 required him to make a trip back to Cambridge. Here he arranged a new home for his mother and sisters (Charlotte was touring with her theatrical company) and straightened out the family affairs. His mother died five years later.

In Urbana during this year Adams began seeing Lucile Wheeler, a teacher of home economics at the university. She was a native of Vermont, a graduate of Mount Holyoke College, to which she was greatly devoted all her life, and had studied dietetics at Columbia. Like Roger, she came from a long-established New England family, being in the tenth American generation descended from Henry

Wheeler, who lived in Salisbury, Massachusetts, about 1640. Her Wheeler forebears moved to Randolph, Vermont, early in the 19th century. She was the daughter of Leonard D. Wheeler (1860–1930) and Jennie A. Smith (1861–1921) and was born in White River Junction, Vermont. Unlike Roger, she was much interested in genealogy and became a member of the Daughters of the American Revolution. She remained an expatriate Yankee for the fifty years she lived in Urbana.

She also went to Washington during World War I to do Red Cross work. There she and Roger continued to meet and they were married on August 29, 1918, at White River Junction, where her father was in business. Roger's financial resources were modest at the time, owing mainly to the financial aid he gave his mother and sisters, and he was obliged to borrow money to get married.[2] Their pictures show a handsome couple; Roger was in uniform that fall and had grown a mustache.

There is a story that Columbia University wrote Lucile Wheeler in Washington, offering her a professorship in household science.[3] She replied that she could not come, as she was now Mrs. Roger Adams. Columbia is alleged to have responded, still hoping that she would accept, "and, as for Mr. Roger Adams, we are confident that we could find something for him to do at Columbia."

When the couple returned to Urbana after World War I, they first lived at 603 West Green Street and in 1924 moved to their large, three-story frame house at 603 Michigan Avenue, where they lived the rest of their lives.[1] They both loved the house, which contained many antiques from Mrs. Adams's Vermont relatives. A spacious living room and a pleasant dining room overlooked a terrace and their well-kept yard and flower garden, and the large library lined with birch bookshelves became both the family living room and the study in which Roger worked nights and Sundays. The five bedrooms and game room easily accommodated the family and the many guests, and upstairs was the sleeping porch where Mrs. Adams liked Roger to sleep—in the fresh air to offset his days spent in the ill-smelling chemistry laboratory.[4,5]

Mrs. Adams was a great cook, and the family shared their holiday dinners with visitors and students and small families far from home. After dinner everyone joined in Ping-Pong and party games. Their daughter, Lucile, Jr. (as Roger called her), remembered those festive days when the graduate students "showed a little girl a lot of attention" and often gave her dolls for her collection. An annual event was the Mother's Day tea to which all the chemistry graduate students and their wives were invited. Mrs. Adams was an accomplished hostess and devoted to Roger; she took great pride in entertaining his guests and in his professional career. Countless American and overseas scientists had happy memories of their hospitality and friendliness.

Mrs. Adams did not have Roger's extroverted personality and she disapproved of liquor, although Roger was a social drinker all his life. He was careful not to offend her with heavy smoking at home. She did not share Roger's pleasure in poker, but they both enjoyed playing bridge. For many years they belonged to an

Illinois University Archives

The Adams Home, 603 Michigan Avenue.

Illinois University Archives

Major Adams (1918)

informal bridge group that met every two weeks, and in spite of his many trips and obligations, Roger missed surprisingly few of these bridge parties.[6,7]

Although Roger lacked the artistic and poetic gifts of his older sister Charlotte, he possessed an irrepressible sense of humor and a talent for light verse, which broke out in entertaining exchanges of doggerel with his family and friends. Each year he penned a valentine in rhyme for his wife, and as Lucile, Jr., grew older, she received one, too.[8]

Adams was never interested in golf, although he played the game, but he was a good tennis player, and in his early years at Illinois he was the champion of the chemistry department. He liked to hike but did so only rarely after moving into Illinois. Popular with local businessmen, he joined the Urbana Rotary Club and became president in 1932.[7]

Early in 1924 Adams had the most serious illness of his active life, a case of empyema that nearly proved fatal before adequate medical treatment brought it under control.[9] His recovery was slow, and he could not work at his usual pace for nearly a year or teach in summer school in G. N. Lewis's department at Berkeley. His old friend Conant took his place at the last minute. Although Adams had a varied educational and professional background, he would have found the Berkeley visit instructive, as Conant did.[10]

During his recovery from his illness, Adams took up stamp collecting, a hobby he pursued with characteristic intensity and success. He specialized in United States commemoratives after 1870 as perfect singles and mint blocks of four, and in his early days of collecting he took time in his travels around the country to visit local post offices and pick different shades of a particular issue. The Urbana postmasters allowed him to choose stamps from their new sheets, and Adams mounted all his selections on handmade paper in leather albums. With typical forthrightness he explained the appeal of his hobby: "What I like about collecting stamps is that it doesn't make me use my brains.... I can fool around with my stamps in an evening and have a good time ... it's just mechanical,... and takes little space.... " He expected to enjoy this avocation in retirement, but his failing eyesight prevented meticulous observation of details, such as shading and perforation. By the time he decided to dispose of it in the 1960s, his collection was very valuable and he sold it in lots to dealers.[11] His interest in stamps was not widely known, and he talked little about it except to other philatelists. Many foreign acquaintances and guests repaid his hospitality and aid by choosing stamps for him and sending issues from their countries.

The story of the Illinois "preps," or, more formally, "Organic Chemical Manufactures," and the evolution of *Organic Syntheses* provide a reasonable starting point for a discussion of Adams's chemical activities at Illinois from 1916 to 1926. C. G. Derick, Adams's predecessor in organic chemistry at Illinois, hired a few graduate students to work during the summer of 1914 preparing organic compounds that would be needed during the next year in teaching and research.[12]

With the outbreak of World War I in August of 1914, this project assumed new significance.

The Eighth International Congress of Applied Chemistry, held in the United States in September 1912, had furnished a striking picture of the accomplishments of the German chemical industry, the results of sustained and cooperative research between industrial and university laboratories. German speakers, exhibits of new synthetics, and a working unit of the newly invented Haber process, producing ammonia from hydrogen and nitrogen for all to see, deepened the impression of German preeminence.[13]

Although the American chemical industry exceeded the German industry in heavy chemicals, it fell behind in specialized organic chemicals, particularly benzene derivatives, which were starting materials for research reagents, dyes, most medicinals, photographic chemicals, and certain analytical reagents. American patent and tariff policy and German cutthroat underselling effectively checked American production. For example, in 1913, firms in this country used 24,000 tons of dyestuffs yet manufactured only 3,000 tons.[14] The superiority of German university research in organic chemistry despite the strides made by American universities intensified the disparity.

Almost completely dependent on German supply houses, every technical, teaching, and research laboratory in the United States underwent an annual ritual, the placement of an order with German suppliers that attempted to anticipate all the needs in fine apparatus, glassware, and chemicals for the coming year. Chemists constantly bewailed the delays in research caused by delays in overseas orders and shipments. As late as the 1930s, university stockrooms still contained bottles of chemicals bought in Germany in an earlier era, but when the Allied Forces resolutely blockaded German exports in 1915, Kahlbaum, Merck, and other German houses were not just six months away; they were completely inaccessible.

In spite of the immediately critical position, the infant American organic chemical industry possessed a solid base and large resources for rapid growth and replacement of German imports.[15] Industrialists, the federal government, and the American Chemical Society mobilized research and production as fast as possible, and the urgent national situation caused an expansion of the Illinois preps labs in 1915.

One of the first workers in the preps labs was E. H. Volwiler; later C. S. Marvel joined the group, and in September 1917 Marvel and another graduate student were given "manufacturing scholarships" instead of teaching assistantships for the academic year. The workers received 25¢ per hour and eventually some academic credit. As every experienced organic chemist knows, the fact that procedures for carrying out a certain reaction have been published in a journal or described in a patent does not guarantee that a repetition of the published procedures in another laboratory will lead to a reasonable yield—or, indeed, to any yield at all of the desired product. Marvel and his associates, therefore, had to do considerable research on conditions and methods to develop efficient and economical procedures for the preparation of key compounds on a sizable laboratory scale. The preps lab workers accomplished prodigious amounts of synthetic work, aided by their ingenuity in improvising equipment out of materials at hand.[16]

The preps lab scheme appealed strongly to Adams: it was work of national importance, it was good chemical training for the students, and it could be made a strong asset to the research program and the national reputation of the Illinois department; in addition, the very strong practical side of Adams's nature found it satisfying. He introduced into the preps operation a systematic scheme of cost accounting for labor and materials, so that the cost per gram of each batch of each compound was calculated. Each worker thus had the cost records of his predecessors on each prep available, and it became a mark of distinction for each worker to cut the cost of each prep. This accounting scheme gave the preps program a valid and impressive financial basis and was excellent training in chemical economics for each participating graduate student, who was obliged to keep the cost records for each of his preps. The preps scheme would have been much less important without this exercise of Adams's business sense, which validated it to the university administration and to outsiders.

Adams wrote a detailed balance sheet covering the period June 1, 1917–February 1, 1918, for the trustees, the figures showing a net profit of $2,800.[17] He recommended that this credit be kept to cover future deficits or overhead costs. A rough balance sheet, covering the years 1917–20, shown below, indicates a surplus of about $3,000. These financial statements by Adams (Table I) show his early fascination with accounting and business matters and foreshadow his expert knowledge in later life in dealing with investments and balance sheets.

Table I. Statement of Organic Chemical Manufactures September 30, 1919

Years	Expenditures	Sales
1917–18	9,256.37	10,299.73
1918–19	12,252.51	10,659.04
1919–20	7,063.95	6,713.26
Inventory of materials (approximate)		3,800.00
Surplus (approximate)	2,899.20	
Totals	31,472.03	31,472.03

The scheme was successful pedagogically, scientifically, and financially. The list of chemicals increased, and 30 university and many technical laboratories bought reagents. The preps lab sold over 60 pounds of dimethylglyoxime, a reagent then used for nickel analysis in steel, and by September 1917 it had prepared about

80 different chemicals.[18] In 1918 about 60 chemicals were in stock and several scores of additional ones were available on order.[19] The *Boston Transcript* carried the story of the Illinois preps lab's contribution to the war effort on its editorial page at the time of the Boston meeting of the American Chemical Society in September 1917,[20] presumably Adams seized the opportunity to present a paper describing them.

Adams's comment in *Science* is revealing. Speaking of the preps lab and possible postwar sources of special organic chemicals, he says, "It is the present intention, at any rate, at the University of Illinois, to continue the work permanently so that . . . there may be a university where a graduate student in organic chemistry may be drilled in commercial methods before he goes permanently into technical work." This, perhaps, was Adams's answer to the numerous editorials and addresses published in *Science* at that time calling for industry to utilize graduate chemists trained in university laboratories for applied research.[21] His business experience in the preps labs and the value of this small operation's products, as well as his observations of wartime expansion of America's factories, must have made him aware of the potential positions for research workers in American organic chemical industry, comparable to those for German chemists. He would also have seen that teaching and research in universities would grow in strength and usefulness with the growth in manufacturing.

The Illinois venture was so valuable and successful that after the war much of the synthetic operation was transferred to the Eastman Kodak Co. in Rochester, New York, under the direction of Hans T. Clarke (later professor of biochemistry at Columbia), who spent several weeks in Urbana studying the operation. This was the origin of the Eastman line of chemicals for research and teaching use. The Illinois work continued in the "summer preps," in which a number of graduate students were paid to prepare chemicals needed for the research program at Illinois and for the course in qualitative organic analysis, providing the laboratory with an unmatched variety of chemicals. In later years the preps labs synthesized amino acids utilized by W. C. Rose of the Illinois department in his classic feeding studies of mice to determine the essential amino acids.

Competition for jobs in the labs continued to be keen over the years, and memories of a summer in preps remained vivid. For a later example: in 1942 H. R. Snyder directed the summer program, and twenty-six workers under two straw bosses labored in two big laboratories throughout the gruelling Illinois heat to fill the orders accumulated during the year. Each worker supervised several simultaneous reactions. Huge, multiliter round-bottom flasks seethed on the benches lining the narrow aisles. Ventilation was meager, perhaps only the hot breeze that blew through the open windows. On many off days when black smoke filled the room, or the student glassblower broke more apparatus than he repaired (in the last weeks his hectic tasks being mainly to seal the stars in the round-bottom flasks), a student needed a special stamina to work at top efficiency to cut the cost of his predecessor's preparation. As the summer wore on, hands so black with chemicals that only a "gunk bath" could clean them, nails that could be "manicured with a cold chisel," and a certain aroma that wafted from each worker became the marks of the preps chemist. Nevertheless, the successful summer produced a set of experienced synthesizers and 200 organic compounds, up by 34 from the summer before.

The tested procedures of the Illinois laboratory were at first published in the *Journal of the American Chemical Society* [22] and later in four *University of Illinois Bulletins* [16,23] on *Organic Chemical Reagents*. Adams considered organizing a group of university laboratories in a cooperative program of preparing needed compounds. This idea proved cumbersome, and, instead, in 1921 Adams, as editor in chief, launched the plan of publishing an annual volume of methods for preparing useful organic compounds with volume 1 of *Organic Syntheses*.[24] Conant, H. T. Clarke, and Oliver Kamm, then at Parke, Davis and Co., were associate editors. This volume contained seventeen procedures, and its preface outlined the plan for the series, calling for an annual publication of preparations and inviting chemists to submit procedures for compounds they had studied in detail. Each preparation was to be checked in a laboratory different from that of the submitter, and exact details of procedures, reagents, yields, physical properties, alternative methods of preparation, and the names of submitters and checkers were to be given. Each volume was to contain about thirty procedures; volume 2 had twenty-five, volume 3, twenty-nine, and thereafter about thirty appeared each year. In the fiftieth volume of *Organic Syntheses* in 1970, Adams described his difficulties in getting a publisher for the first issue because all the publishers he initially approached used the same referee, who repeated his negative recommendation.[25] John Wiley and Sons finally agreed to publish the first volume and had no reason to regret it.

Many organic chemists cooperated in submitting preparations, and they regarded service on the editorial board as an honor. The success and continued vitality of *Organic Syntheses* in its sixth decade shows that it filled a real need. The annual redbacked volumes and the five green collective volumes, containing the carefully edited contents of the nine previous annual issues instead of the five proposed in the first preface, are a remarkable record of the development of synthetic organic chemistry since 1921. New reactions, new reagents, new instrumentation, new ideas on reaction mechanisms, new editors, and new submitters have kept the series an active force in organic chemistry. The first volume, modestly styled a "pamphlet" in the preface, and its many successors are a fitting monument to Adams, his associates, their successors, and to the Illinois preps labs. In a letter written for Adams's retirement dinner in 1954, Conant recalled the discussion of the idea of *Organic Syntheses* in 1918 during their war service in Washington, and noted that it should be called the "Adams Annual" because he was the moving spirit.[26] This series was close to Adams's heart, and he took an active part in it up to his last years. *Organic Syntheses* was the model for *Inorganic Syntheses* and *Biochemical Preparations*, which have been useful in their fields.

The aggressive development of the preps program first brought him into national prominence in the chemical world. He undertook a vigorous letter-writing program to leading universities telling them of the availability of the Illinois chemicals.[27] He also distributed bound volumes of reprints of publications from the organic division to a large mailing list of leading chemists in this country and abroad, and he sent some complimentary copies of *Organic Syntheses* overseas. The impact of these publications in making the Illinois department recognized worldwide was very great, as shown by many letters, including the one below of December 7, 1927, from E. P. Kohler:[28]

> The bound publications of your Division of Chemistry have arrived and I am grateful for them. I do not agree with your statement that "unfortunately a little advertising is necessary." I do not believe that the Department of Chemistry of the University of Illinois now needs any advertising in this country. However, so long as the advertising takes such a useful form I do not much care what name you give to it.
>
> I should think that you would be somewhat reluctant to issue a volume such as that which represents your publications for the years 1922 and 1923 because I doubt that you will ever again be able to produce such a volume of papers in so short a time. In fact, I am incapable of understanding how you could survive so much writing!
>
> I have had a letter from Fuson which delighted me because he evidently is very happy with you and likes the ways in which you do things, thinks much more highly of all your ways than Carothers ever thought of ours, and certainly writes with much greater enthusiasm. He ought to get on well with you.

Literature Cited

1. *University of Illinois Directory*, published by the University, October 1916, p. 25. Later residence: *Illinois Faculty and Student Directories.*
2. Private communication from Lucile Adams Brink. Adams's continued help to his family is shown by canceled checks after his marriage. The genealogy of Lucile Wheeler is taken from family documents loaned to us by L. A. Brink.
3. E. H. Volwiler to DST, August 21, 1978.
4. Private communication from L. A. Brink.
5. RAA, 16, Personal, Family, 1963–1967; RA to J. W. Peltason, December 15, 1966, when Adams was offering the house for rent.
6. Conversation in 1977 with Mrs. William Palmer of Urbana, a neighbor and close friend of the Adamses for over fifty years.
7. Conversation with W. C. Rose, January 10, 1977.
8. RAA, 5, 1942–1949; also Ref. 4.
9. RAA, 4, 1924; contains many letters from Mrs. Adams to her family and to Roger's sister, Mary, about Roger's illness.
10. Conant, pp. 64–6.
11. RAA contains only a few letters to stamp dealers, considering the size of his collection;

John R. Johnson to DST, November 20, 1977. In RAA many personal letters refer to foreign stamps. A description of Adams's collection is by Tom Bird, *The Palm of Alpha Tau Omega*, April 1939, pp. 8–10; also Ref. 4.
12. R. L. and R. H. Shriner, *Cumulative Indices, Organic Syntheses*, Wiley, New York, 1976, pp. 423–27; a brief account of the preps labs and *Organic Syntheses*.
13. C. A. Browne and M. E. Weeks, *A History of the American Chemical Society*, American Chemical Society, Washington, D.C., 1952, p. 107.
14. L. F. Haber, *The Chemical Industry, 1900–1930*, Oxford Press, Oxford 1971, pp. 108–34, 173–83. Also W. Haynes, *American Chemical Industry*, Vol. II, Van Nostrand, New York 1945, pp. xiii–xiv, 3–4.
15. A. J. Ihde, *J. Chem. Educ.*, *53*, 741 (1976). Our own extended studies of pre-1914 American organic chemistry support Ihde's conclusion.
16. C. S. Marvel to DST, January 23, 1976; taped interview with Marvel by DST, March 1, 1977.
17. Statement, signed by RA, Illinois Archives, Dean's Office, Department and Subject File, 15/1/1, 17, chemistry; undated but after February 1, 1918. The statement starts by saying that organic chemical manufacturing started in June 1917; this must mean that outside sales started then, when Adams took charge of the project. The start of the preps project by Derick in 1914 is beyond question.
18. RA, *Science*, *47*, 225 (1918); *J. Ind. Eng. Chem.*, *9*, 685 (1917); RA's description of the origin and growth of the preps labs.
19. RA, *J. Am. Chem. Soc.*, *40*, 869 (1918).
20. RAA, 57, Press Clippings, 1917–20. For description in 1942: RAA 10, 1940–1951; *The Zeta Ion*, *8*, n.s.1, p. 3 (Fall 1942–43).
21. W. R. Whitney, "Research as a National Duty," *Science*, *43*, 629 (1916), and many others 1914 on.
22. RA and Oliver Kamm, *J. Am. Chem. Soc.*, *40*, 1281 (1918); RA, Kamm and C. S. Marvel, ibid., *41*, 276, 789, (1919); *42*, 299, 310 (1920).
23. The first was RA, Kamm, and Marvel, *University of Illinois Bulletin*, XVI, no. 43; reviewed in *J. Am. Chem. Soc.*, *42*, 1074 (1920) by Moses Gomberg with complimentary remarks on the value of the Illinois preps program. Gomberg reviewed the second issue, ibid., *43*, 1748 (1921). The four bulletins were dated June 23, 1919; October 11, 1920; October 9, 1921; and October 23, 1922 (information from the Du Pont Lavoisier Library).
24. *Organic Syntheses*, Vol. 1, Wiley, New York, 1921.
25. RA, *Organic Syntheses*, *50*, vii (1970).
26. RAA, 21, Conant; J. G. Conant to R. M. Joyce, July 13, 1954; Conant, p. 50.
27. RAA, 4, 1915–1916; a handwritten note from G. N. Lewis of Berkeley of December 15, 1916, acknowledges RA's letter to him.
28. RAA, 7, E. P. Kohler.

In World War Service, 1917–18

The National Research Council (NRC), a group of scientists sponsored by the National Academy of Sciences, appointed numerous committees to study problems caused by the war. Adams served in 1917 on two subcommittees of the Chemistry Committee of the NRC. The Subcommittee on Organic Chemistry consisted of William A. Noyes and Adams. A report of the Chemistry Committee for November 1, 1917, describes their most important activities as an extension of the Organic Chemical Manufactures at Illinois:[1]

> ORGANIC CHEMISTRY ... The most important undertaking by this Committee has been the organizing of various university and research laboratories by Dr. Roger Adams for the production of special organic chemicals, the need of which is urgent, but yet which are required in such small amounts relatively as to make it difficult to induce any commercial manufacturers to engage in their production. Many of these materials are of the utmost importance for the prosecution of research and analytical work, and Dr. Adams is now prepared to supply nearly 100 different compounds of this character, which it is difficult or impossible to secure elsewhere. The list is being steadily added to, and the value of this service is being increasingly appreciated by the chemists throughout the country.

The second subcommittee, on the chemistry of synthetic drugs, chaired by Julius Stieglitz of Chicago, had three members, including Adams. It was the duty of this subcommittee to recommend government licenses for pharmaceutical houses permitting the manufacture of certain needed drugs. When Adams reached Illinois in 1916, he started work on the synthesis of local anesthetics and became a consultant for Abbott Laboratories. The facts that Adams was connected with Abbott, was doing research on drugs, and had a patent in this field caused complaints by other pharmaceutical houses about a possible conflict of interest in his case. Steiglitz wrote to Adams pointing this out and requested Adams's resignation, disclaiming any doubts of Adams's integrity.[2] Adams did resign from the subcommittee and also wrote to Burdick of Abbott, offering to resign as consultant if the connection embarrassed Abbott. The Abbott people saw no reason for him to do so.

Adams's further service in World War I was in research on war gases. The Germans first used gas warfare in World War I on April 22, 1915, at Ypres, and a favorable wind carried clouds of chlorine gas through the British lines. The British troops were entirely without protection and the results were devastating. Research began at once in Britain and France, and improvised gas masks soon gave protection against chlorine. When the United States entered the war in April 1917, the U.S. Army had no gas masks, no provisions for offensive gas warfare, and no research program on the problems involved.[3,4]

In the spring of 1917 Van H. Manning, who was director of the Bureau of

Mines at Pittsburgh and familiar with gas masks in connection with toxic gases in mines, suggested to the military committee of the NRC that a research program should be started at Pittsburgh on testing gas masks. The NRC accepted this offer and on April 3, 1917, set up a Subcommittee on Noxious Gases with Manning as chairman. Manning appointed George A. Burrell, who was also experienced with gas masks, to direct the research at the Bureau of Mines, and Burrell's staff included members of the bureau's research group in addition to three men to become very well known in later years: Yandell Henderson, a physiologist from Yale; Bradley Dewey, an industrial research chemist; and Warren K. Lewis, assistant professor of chemical engineering at MIT. The Bureau of Mines procured 25,000 gas masks, which were protective against chlorine but offered no protection against chloropicrin, which the Germans had recently introduced.

Meanwhile, Charles L. Parsons, chief chemist of the Bureau of Mines and executive secretary of the American Chemical Society, surveyed American chemists and engineers to determine their numbers, experience, and location. The society urged its members not to enlist in the army but to await assignment to technical work. Manning's Subcommittee on Noxious Gases called upon numerous university research groups to work on chemical warfare problems, among the leaders being E. Emmett Reid, Remsen's successor as organic chemist at Johns Hopkins.[5] Reid asked several university organic chemists, many of them Hopkins Ph.D.'s, to submit compounds for testing as war gases. His request brought in fifty-six compounds for testing.

In spite of the hard work at the Bureau of Mines and at numerous branch laboratories set up in universities with facilities for research, it was apparent by the end of the summer of 1917 that the rather loose-jointed voluntary research effort was not very efficient. Negotiations were completed during the summer for a central research laboratory on chemical warfare problems at the American University in Washington, D.C., to be supported by the army and the navy and administered by the Bureau of Mines. Eventually sixty buildings arose on the ninety-two acre campus, which had contained originally only two buildings. J. B. Conant led one of the first groups and started work on organic chemical problems on September 22, 1917, in laboratories that were not yet complete. The personnel were partly civilian and partly military, serving in various branches of the army. After considerable discussion and against the views of the American Chemical Society, as expressed in an editorial in the *Journal of Industrial and Engineering Chemistry*, President Wilson transferred the work at American University in June 1918 to the research division of the newly created Chemical Warfare Service of the army. By the end of the war in November 1918, the research division at American University had a total of about 1,700 people, of whom roughly 1,000 were technical, and of these, eighty percent were military with over 100 commissioned officers.

By the summer of 1917 the expressed policy of the army was to draft all men under age thirty-five, and in preparation for this, following army regulations,

Adams and students drilled conscientiously in the Illinois Armory in the winter of 1916–17, using wooden guns.[6] Adams was not drafted, but he was involved early in the research in war gases; according to his accounts written many years later, he was in Washington "from the fall of 1917 until after the war ended in 1918."[7] This agrees with the recollection of E. H. Volwiler that Adams spent much of this autumn away from Illinois when Volwiler was completing his Ph.D. thesis.[8]

On February 23, 1918, Adams was appointed by the Department of the Interior as "Gas Chemist in the Bureau of Mines, at a salary of $3,600 per annum, effective on the date of entrance on duty." He arranged to leave Illinois in March, and President James granted him "absence without pay" from March 5, 1918, but allowed him $40 per month to September 1918 for his continuing university work directing the preps laboratory and his research students.[9]

As Adams wrote after the war, "The work was later transferred to the War Department July 1, at which time I was put in charge of the Organic Chemical Unit of the Offense Chemical Research Section of the Chemical Warfare Service. On September 14, I received the commission of Major in Chemical Warfare Service and was discharged December 23, 1918."[10] The recommendations for the commission came from both Lt. Col. Arthur B. Lamb, C. W. S., and Kohler, who was serving as the civilian head of the section, and Kohler's statements make it clear that it was not a promotion from lower military rank but from civilian status.[11] Thus Adams was in uniform for only a few months, unlike Conant who served in the lower ranks before becoming a major.[12]

Adams described his experiences at American University as an exciting period for chemists.[13] Those under 35 were drafted and the older ones who could be spared from "the embryonic chemical industry" received responsible posts as army officers or civilians.

> At the beginning the chemists were required to have military drill every afternoon at 4 and that caused complications, for some of the poorer chemists had the higher ranks and conducted the military training accordingly. But in the laboratories responsibilities had to rest on those who were best trained in chemistry—not in the army procedures. Fortunately it was possible after some time to have this regulation for drill rescinded, much to the delight of the chemists.
>
> There were three offense chemical research laboratories—one headed by J. B. Conant, later president of Harvard University and Ambassador to Germany; one directed by W. Lee Lewis, a professor of organic chemistry at Northwestern University; and a third by myself. In my lab one of the best group leaders was Henry Gilman, recently retired professor at Iowa State. He was a private and frequently supervised men of the noncommissioned rank as well as officers as high as captain. My laboratory and Lewis' laboratory were devoted primarily to the study of arsenic compounds, lacrymators, and miscellaneous gases; Conant's laboratory, almost entirely to mustard gas. The ventilation in the laboratories was very ineffective, and many of the chemists became sensitive to the

poisoning. Conant and I had a good system; he would switch to my laboratory chemists who became very sensitive to mustard, and I would switch to his those badly poisoned with arsenic. I can assure you the morale was low by December, 1918, and nobody was more exuberant at war's end than the chemists.

The Chemical Warfare Service contributed something of importance to chemistry in the United States; it brought together 80% of the chemists. Lasting friendships were made—many of them life long. And the American Chemical Society meetings thereafter were very much better attended for they presented the opportunity for chemical friends meeting chemical friends again. This post-war period marked the beginning of the ever increasing success and effectiveness of the Society.

In spite of the improvised research facilities at American University, the morale of the people there seems to have been high; their monthly newspaper described recreational activities such as dances, a glee club, and baseball and basketball leagues.[14] Conant mentions the friendly informal contacts among organic chemists and notes that Adams was the leader in this group. At American University the two friends first considered the idea of publishing what later became *Organic Syntheses*.[12]

As the war continued, the Germans followed chlorine gas with phosgene and its dimer; these were asphyxiant gases. Later they fired gas shells containing chloropicrin, a sternutator ("sneeze gas") that forced the soldier to throw off his gas mask, and in July 1917 they attacked with mustard gas, a toxic vesicant, or blister producer. A high-boiling liquid, mustard did not evaporate readily, and this persistent gas caused many casualties among allied troops.

$COCl_2$ phosgene

$Cl_3CCOOCl$ phosgene dimer

Cl_3CNO_2 chloropicrin

$ClCH_2CH_2SCH_2CH_2Cl$ mustard

American preparations to retaliate led to intensive study of the German gases and to the synthesis of new compounds. W. Lee Lewis synthesized lewisite by a reaction studied many years earlier by J. A. Nieuwland (1878–1936), later a professor at Notre Dame.

$$HC{\equiv}CH + AsCl_3 \xrightarrow{AlCl_3} ClCH{=}CHAsCl_2$$

lewisite

Conant's unit was sent to Willoughby, Ohio, in June 1918 to build a closely guarded plant to produce this new gas.

Adams's unit also worked extensively on arsenicals and developed the sternutator adamsite; it was readily prepared and Adams's group synthesized 100 pounds.

Diphenylamine (NH) $\xrightarrow{AsCl_3}$ (As–Cl, NH)

adamsite

Adamsite was a solid, but if dispersed in finely divided form, it would penetrate charcoal gas masks. The Allies did not use the arsenical gases in the field in World War I but did study their properties and methods of dispersal in detail.

Adams's group also worked out a simple method of preparing chloracetophenone, found by Reid to be a very effective lacrymator. Although not used in the war, it is the tear gas commonly used today by police.

Benzene + $ClCOCH_2Cl$ $\xrightarrow{AlCl_3}$ $C_6H_5COCH_2Cl$

chloracetophenone

The methods of preparing chloracetic acid, its acid chloride, and chloracetophenone as worked out by Adams's group served for the large-scale preparation of this material.[15]

Adams's experience with the chemical warfare research program was far more important for his future career than would be predicted from the relatively short time involved. It increased his knowledge of chemistry, particularly the problems related to scaling up laboratory preparations, and enhanced his scientific standing. He obtained administrative experience in supervising a large group of workers and in army procedures, and he renewed old friendships and made many new friends. He had made a contribution to the national scientific effort, but it is extremely doubtful that either he or Conant dreamed that within slightly more than two decades they would both be back in Washington engaged in war research on problems far more varied and fateful for the nation's future than those of 1917–18.[11] The experience in World War I was invaluable to them both in 1940–45.

Adams returned to Illinois more anxious than ever to develop his own research program in the freedom of a university department. In a very short time his national reputation as an extremely promising and able chemist brought him several tempting offers of other academic appointments and of positions in industrial research laboratories. His friends uniformly advised him to stay at Illinois. The Illinois administration now realized that they were dealing with an exceptional man and in 1919 made him a full professor with tenure at $4,000 per year. In the chemistry department he became head of the organic division.[16] For comparison, salaries for full professors at Harvard were $4,000–5,500 in 1918–19 and were raised the following year to $6,000–8,000 as the result of an endowment drive.[17]

A letter from C. L. Jackson in Boston, January 3, 1919, about one of Adams's offers says:[16]

> When you told me about the Goodrich place, I did not say half what I wanted to, as I felt I did not know the conditions, but afterward I wished I had said what follows and so send it to you, even if I am too late.
>
> I think your leaving pure research for industrial chemistry would be a very serious injury to science. You are already prominent in research and in your excellent plans, and certain to grow more and more important. If all our good research men and teachers go into technical work, where are we going to get well-educated chemists we need in the immediate future? In other words I am sure you will be more useful as a teacher than as the chemist of the most important factory. I should never have accepted any industrial offer no matter how tempting, but then I have never been in want of money. . . . How I did enjoy your call!

Although Jackson may have had a rather cloistered view of what "the chemist" in a "factory" actually did, there is no question of his devotion to Adams and his career. Adams himself wrote in 1934, "Perhaps the greatest problem in my life, which has arisen several times, has been to decide whether to leave teaching for industry. My decision to remain in academic work and scientific research is not regretted and it is very probable that it will always be a university position for me from now on. The satisfaction of seeing former students succeed is greater than the business world can realize".[18]

His commitment to Illinois as his home and to the progress of the university is shown by the following letter of January 22, 1925, from Adams to the editor of the *Harvard Alumni Bulletin*[19]:

> Under the caption "The State Universities" an interesting editorial in the Alumni Bulletin of January 15th discussed the matter of increased professorial salaries in certain Middle Western institutions. It was pointed out how these

institutions had developed in recent years and how their prestige was becoming a factor that the eastern schools must consider. The editorial would impress one as being written by a well informed individual who knows the colleges of the East and West, were it not for one phrase which makes it seem obvious that the writer is a New Englander imbued with the superiority of the East. The phrase follows: "In some of the Western institutions ... the facilities for research in some fields are almost as good (as at Harvard)." Until I went to instruct in the Middle West I had lived most of my life in the vicinity of Cambridge and felt as the writer does about the inferiority of anything away from Harvard. Now, after almost nine years in the Middle West I have become enlightened as to the facilities, especially in science and engineering, in many of the Middle Western Universities. I would like to have the writer of the editorial and others who think the same way come out and get acquainted.

Since the time is near when conditions are to be remedied it is almost unfair to select facilities in chemistry at Harvard for comparison with those in other places. I feel it is permissible, however, since I am a chemist who did all his undergraduate and graduate work there. Some of my best students I urge to go to Cambridge and to continue their study under some of the eminent Harvard professors. But after such recommendations to the student I add an apology for the laboratory and explain that they must not expect there the facilities in many lines of chemistry that they find here. In addition, I urge upon them the great value gained from different contacts and the enlarged horizon which comes from a knowledge of different institutions. The East has much to offer students from the Middle West, and the West can open the eyes of the East surprisingly in many ways. As football well-played is no longer confined to the Big Three, so in educational facilities the Big Three have their competitors.

The writer of the editorial for accuracy's sake might have said, "In some of the Middle-Western institutions, ... the facilities for research in some fields are almost as good, in many as good and in some better."

His advancement and recognition now secure at Illinois after the Washington interlude, Adams turned to his research and teaching programs with zeal.

Literature Cited

1. Report in the National Academy of Sciences Archives.
2. RAA, 4, 1918–1919; Steiglitz to RA, June 19, 1918.
3. The American war research effort from the standpoint of the American Chemical Society is described in C. A. Browne and Mary E. Weeks, *A History of the American Chemical Society*, American Chemical Society, Washington, D.C., 1952, pp. 108–26.
4. The research on chemical warfare is covered in detail by D. P. Jones, *The Role of Chemists in Research on War Gases During World War I*, Ph.D. Thesis, University of Wisconsin, 1969, directed by A. J. Ihde; University Microfilms 69–22406. Our account is based on Jones, unless otherwise indicated. G. A. Burrell, *J. Ind. Eng. Chem.*, *11*, 93 (1919), gives a contemporary survey of the whole program.

5. E. Emmett Reid, *My First Hundred Years*, Chemical Publishing Co., New York, 1972, pp. 120 ff.; Jones, op. cit., p. 108.
6. E. H. Volwiler to DST, August 21, 1978.
7. RAA, 36, J; RA to D. P. Jones, April 7, 1969.
8. E. H. Volwiler to DST, June 10, 1979.
9. RAA, 64, 1913–64.
10. University of Illinois Archives 4/5/50, 4, 1, University Senate Committee on the History of the Participation of the University in World War, 1915–1923, Chemistry. Another account by Adams with minor differences is in *Harvard College Class of 1909, Tenth Anniversary Report*, Boston, 1920, p. 1. These successive class reports together contain the closest thing to a personal autobiography that Adams ever wrote.
11. RAA, 4, 1918–1919.
12. Conant, pp. 49–50.
13. RAA, 4, Noller Symposium; lecture at the retirement dinner for Carl R. Noller at Stanford, May 20, 1966. Some of this material is also covered in the letter to Jones.
14. Ref. 4, Jones, p. 122.
15. RAA, 4, 1918–1919, and Chemical Warfare Service.
16. RAA, 4, 1918–1919, 1920–1922.
17. S. E. Morison, *The Development of Harvard University, 1869–1929*, Harvard University Press, Cambridge, 1930, p. xli, footnote.
18. *Harvard College, Class of 1909, 25th Anniversary Report, 1909–1934*, Harvard University Press, Cambridge, pp. 2–3.
19. RAA, 4, 1925–1929.

Scientific Work, 1916–26

Some technical discussion of a portion of Adams's scientific papers, which number 425 and include the research of 184 Ph.D.'s, over 50 postdoctorates, and numerous master's and bachelor's candidates, reveals the scope and significance of his ideas and work. To omit this evaluation would be completely misleading. Interesting and important as his activities outside Urbana became, they rested on the solid scientific base of his chemical publications.

When he arrived in Urbana in 1916, Adams, in collaboration with Oliver Kamm, a Noyes Ph.D. and instructor in chemistry, undertook the preparation of the local anesthetic procaine and the sedative barbital, which were no longer available from Germany and were badly needed. Adams and Kamm developed procedures suitable for large-scale preparation of procaine and carried on a lively correspondence with A. S. Burdick, the medical director, and W. C. Abbott, the founder, of Abbott Laboratories. This work resulted in a patent on procaine synthesis and was followed later by a series of papers and patents on local anesthetics, including an analog, butacaine, of procaine.[1]

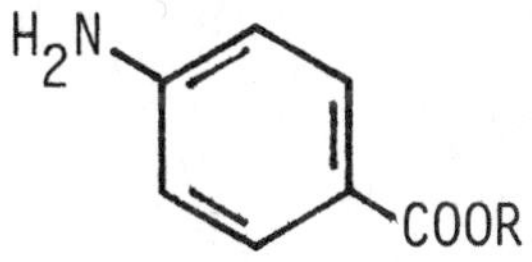

H_2N — (benzene ring) — COOR

procaine, R = $-(CH_2)_2N(C_2H_5)_2$

butacaine, R = $-(CH_2)_3N(C_4H_9)_2$

This early work on compounds with biological activity started Adams's lifelong interest in this general field, and it also led to his becoming a chemical consultant to Abbott Laboratories in 1917 at a retainer of $50 per month.[2] This connection continued until the 1960s and was strengthened by the fact that Adams's first Ph.D. student, Ernest H. Volwiler, joined Abbott as research chemist in 1918. Volwiler had taken a master's degree with Derick at Illinois before Adams's arrival and was considering an industrial position when Paul Anders, the department glassblower and a notable character in his own right, persuaded him that he would be foolish to leave without a Ph.D.[3] Volwiler became president and chairman of the board at Abbott,[4] and the Volwilers remained among the Adamses' closest personal friends.

Adams's own Ph.D. thesis on alkali-insoluble phenols had been published only in preliminary form in 1910 after H. A. Torrey's death. At Illinois Adams did additional work on the problem and published a long paper by himself.[5] This paper was rather anticlimactic, because it simply showed that phenols that failed to dissolve in 10% aqueous sodium hydroxide behaved this way because they were too slightly soluble in plain water. More sophisticated tests for solubility and acidity were available even then. The paper illustrates a chemical limitation of Adams; he was superb in doing structural determinations and syntheses but was never really at home dealing with the applications of physical chemistry to organic chemical problems. This weakness was to be reflected to some extent in the development of the Illinois department.

A series of papers described reactions of oxalyl chloride. Henry Gilman, working as an undergraduate at Harvard with Adams, had found that oxalate esters were easily obtained from oxalyl chloride and phenols if pyridine was present.[6] Weeks[6] found that primary alcohols gave oxalate esters, secondary alcohols dehydrated, and tertiary alcohols formed the corresponding chlorides. These and later observations underlined the difference in reaction paths among primary, secondary, and tertiary alcohols, already noted by J. F. Norris and others.

At Illinois Adams found that oxalyl chloride formed the acid anhydrides from carboxylic acids; in some cases intermediate mixed anhydrides could be isolated, which gave the symmetrical acid anhydrides on heating.[7] Oxalyl chloride and the salts of acids were found to yield the acid chlorides, a very mild and useful synthetic procedure.[7]

$$2\ C_6H_5COOH + ClC(=O){-}C(=O)Cl \longrightarrow C_6H_5C(=O){-}O{-}C(=O){-}C(=O){-}O{-}C(=O)C_6H_5 + 2\ HCl$$

$$\downarrow \text{heat}$$

$$C_6H_5C(=O){-}O{-}C(=O)C_6H_5 + CO + CO_2$$

Volwiler[8] showed that benzoyl bromide and benzaldehyde gave a bromo ester, which could be converted to a variety of other compounds. With oxalyl bromide and benzaldehyde, similar products were formed: $C_6H_5CHBrOCOCOOCHBrC_6H_5$ and derivatives. The Volwiler compounds, which were esters of α-haloalcohols, were stable in many cases in the presence of tertiary bases such as pyridine.

$$C_6H_5COBr + C_6H_5CHO \underset{H_2O}{\rightleftharpoons} C_6H_5COOCHBrC_6H_5$$

$$\xrightarrow{C_6H_5COOAg} (C_6H_5COO)_2CHC_6H_5$$

His experiences with arsenic compounds during the chemical warfare work led Adams to examine the synthesis of several types of arsenic derivatives; the hope that some of them might show useful medicinal properties, like Ehrlich's salvarsan, was not realized.[9] There are many American papers after World War I on synthesis and analysis of salvarsan (arsphenamine) and related compounds, but little of medicinal interest resulted. Tests of many of Adams's compounds against trypanosomes, which cause African sleeping sickness (trypanosomiasis), did not show promising activity.

An elegant piece of chemical reasoning and experimental work was the determination of the structure of disalicylaldehyde as that below:[10]

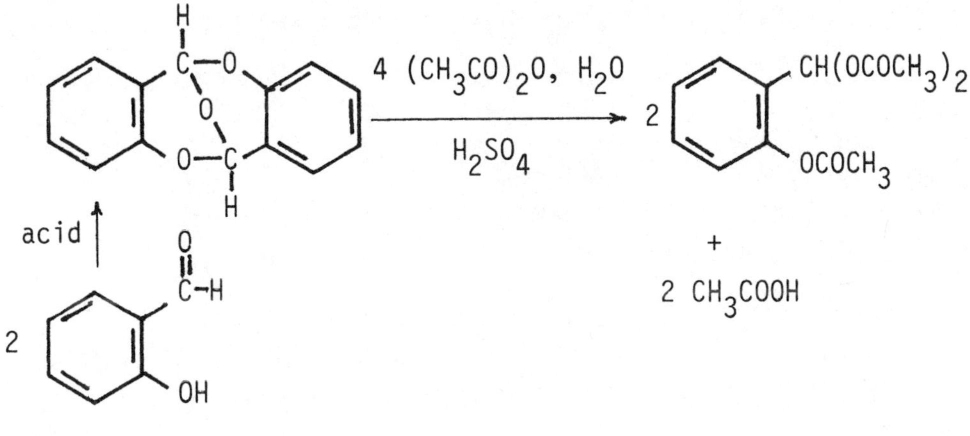

$$2\ \text{salicylaldehyde} \xrightarrow{\text{acid}} \text{disalicylaldehyde} \xrightarrow[H_2SO_4]{4\ (CH_3CO)_2O,\ H_2O} 2\ C_6H_4(CH(OCOCH_3)_2)(OCOCH_3) + 2\ CH_3COOH$$

salicylaldehyde

This structure is compatible with absence of carbonyl and hydroxyl groups, with the hydrolysis by acid but not by base, and with the action of acetic anhydride. Support by analogy for the structure was given by showing that saligenin and benzaldehyde condensed to form a similar structure.

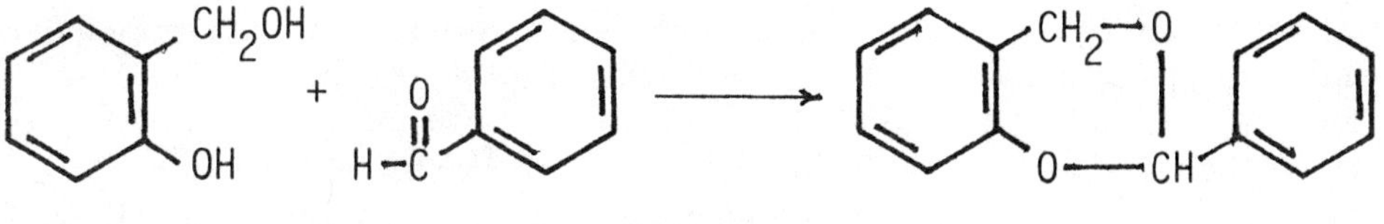

$$C_6H_4(CH_2OH)(OH) + H{-}C(=O){-}C_6H_5 \longrightarrow C_6H_4(CH_2{-}O)(O{-}CH{-}C_6H_5)$$

saligenin

Similar acuteness in establishing the correct structure for a compound for which several alternatives had been considered was shown in the assignment of the structure below to dehydroacetic acid.[11] This paper is notable both for the classic economy of experimental work with which the structure was proved and for the classic economy of style in which the results were presented.

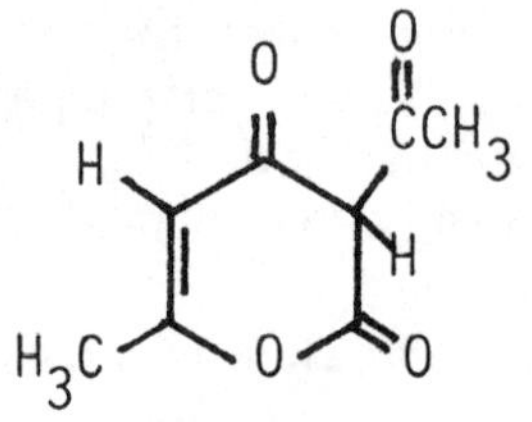

dehydroacetic acid

A valuable experimental contribution was the development of a simplified method for preparing aromatic hydroxyaldehydes; the use of the dangerous anhydrous hydrogen cyanide in the original Gatterman procedure was avoided by using zinc cyanide.[12] This is a general procedure.

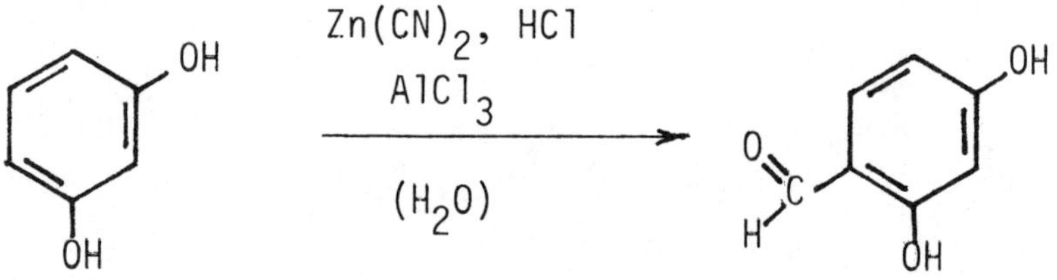

$$\text{resorcinol} \xrightarrow[(H_2O)]{Zn(CN)_2,\ HCl,\ AlCl_3} \text{2,4-dihydroxybenzaldehyde}$$

One of Adams's triumphs in research during this period was his discovery in 1922 of "Adams platinum oxide" for catalytic reduction of multiple bonds. More than any other single piece of work, this produced an extraordinarily important reagent, which is used and associated with Adams's name wherever organic chemistry is pursued.[13]

Ever since the turn of the century, when Sabatier found that finely divided nickel catalyzed the addition of hydrogen to unsaturated groups, the development of good catalysts for hydrogenation had been an important scientific and practical problem. The hydrogenation, or "hardening," of liquid vegetable oils to solid fats for soap and shortening became an important industry. Finely divided noble metal catalysts (platinum and palladium) were prepared in various ways for catalytic reduction on a laboratory scale. One of the leaders in this field was Richard Willstätter. The varied ways proposed for preparing these catalysts and the frequent reports of unreproducible results show how unsatisfactory these platinum and palladium catalysts really were. Adams discovered a simple and reproducible way of making a platinum catalyst of very wide applicability for catalytic reduction, which at once superseded most earlier noble metal catalyst preparations.

There was a large element of serendipity in the discovery of Adams platinum, and the story is best told in his own words:[14]

> A problem of more general significance [than the structure of dehydroacetic acid] was being investigated simultaneously. It was a search for a platinum catalyst to be used in the reduction of organic compounds. Many platinum catalysts had been described previously but it was seldom possible to repeat a preparation and to obtain a product of similar or high activity. A year or two of tedious experimentation to improve what appeared to be the best previous directions as described by Willstäter and Waldschmidt–Leitz resulted in defining precise details for the preparation of a catalyst of uniform and high activity. As a consequence, an ounce of chloroplatinic acid was employed to reproduce a large amount of catalyst at one time. The operation involved the reduction of chloroplatinic acid to platinum black by means of formaldehyde and alkali. After the precipitation and before the platinum had been isolated in this large run an accident happened. The casserole broke and the platinum black was strewn over an old wooden table top. It was finally scraped up along with splinters, wood surfacing material and other debris. The normal procedure for recovering the platinum was not adequate to remove all the impurities and hence it was decided to resort to a fusion with sodium nitrate to burn out the residual organic material. In this process a brown platinum oxide formed which was found to be reduced readily in the presence of hydrogen to a highly active platinum black catalyst. This is still the most active and most readily prepared platinum catalyst for hydrogenation reactions and the most widely used in research laboratories. In a series of papers which followed, its use in the reduction of various types of organic compounds was explored.

The equation for the fusion reaction was written

$$H_2PtCl_6 + 6NaNO_3 \longrightarrow PtO_2 + 4NO_2 + O_2 + 2HNO_3 + 6NaCl$$

The optimum conditions for preparing the platinum oxide, its usefulness in reducing many types of unsaturated compounds, including aromatic rings, and the promoting or inhibiting effects of various salts and acids were systematically studied by Adams in a series of eighteen papers with a dozen collaborators, among whom were W. H. Carothers and R. L. Shriner. These researchers found, for example, that ferrous salts promoted the reduction of aldehydes to alcohols but they inhibited the reduction of carbon–carbon bonds. Thus unsaturated aldehydes could be reduced selectively to unsaturated alcohols.[15]

Adams's discovery of his platinum catalyst was reported in the usual way in the scientific literature, without any mention of its serendipitous character.[16] The apparatus developed for its use at one to three atmospheres of hydrogen was later manufactured by a company run by Adams's colleague, Samuel G. Parr. The Parr shaker and Adams platinum became at once a standard combination in most organic chemical research laboratories. Other types of platinum catalyst became obsolete,[17] and a recent monograph lists Adams's name over sixty times.[18] Adams had become the best-known structural organic chemist in the United States by 1925, at the age of thirty-six.

His scientific reputation was enhanced by two other important series of research papers. He was interested throughout his career in the chemistry of naturally occurring compounds, and he developed syntheses for various naturally occurring anthraquinones, similar in structure to the valuable dye alizarin. Adams synthesized[19] a similar natural product, emodin, which was also used as a dye; his synthesis of emodin followed a fairly standard course but with some original additions. He later developed a new route to this series, using as starting material opianic acid, an aldehyde acid obtainable from the opium alkaloids. The application of this ingenious scheme to the synthesis of morindone is shown in Scheme 1.[20]

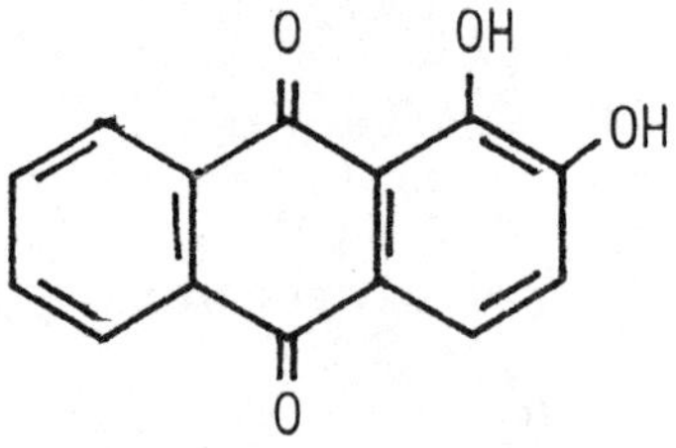

alizarin

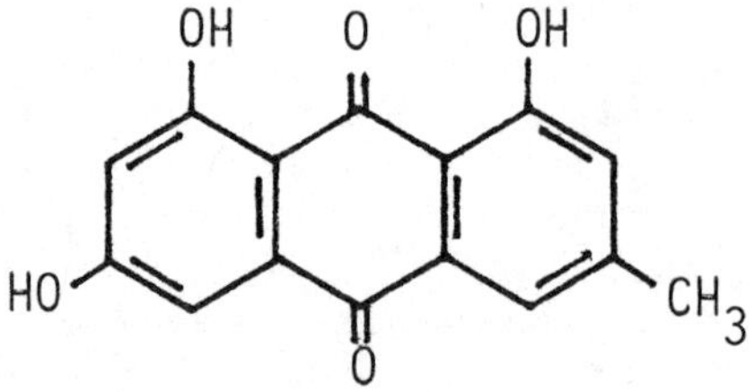

emodin

In all these syntheses the reliability of the methods and the structures of the intermediates were conclusively established.

The work of Adams's group in this period that became most famous, however, second only to the research on the platinum catalyst, was undoubtedly the structural elucidation of chaulmoogric and hydnocarpic acids and the syntheses in this field. These acids were constituents of chaulmoogra oil, a natural oil that had been used with some success in treating leprosy. The conclusive elucidation of the structures of these compounds[21] required repetition and critical reinterpretation of older experimental work by others, which ruled out some suggested structures. Reduction with Adams platinum gave the dihydro acids, (see Scheme 2), and these were synthesized by a general method,[22] using an aldehyde obtained by ozonizing unsaturated esters (see Scheme 3). The product was identical with natural dihydrochaulmoogric acid.

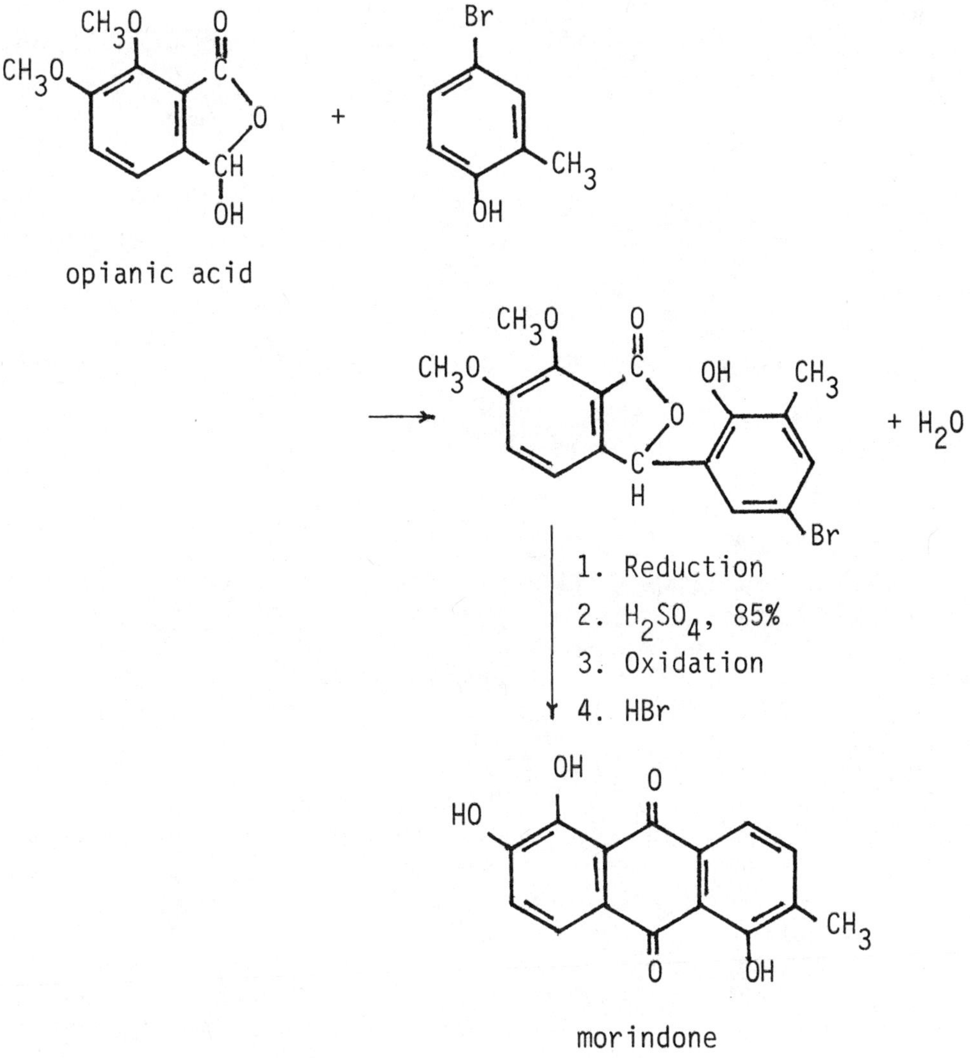

CH_3O CH_3O O C O CH OH + Br CH_3 OH

opianic acid

→ CH_3O CH_3O O C O C H OH CH_3 Br + H_2O

1. Reduction
2. H_2SO_4, 85%
3. Oxidation
4. HBr

OH HO O O CH_3 OH

morindone

Scheme 1

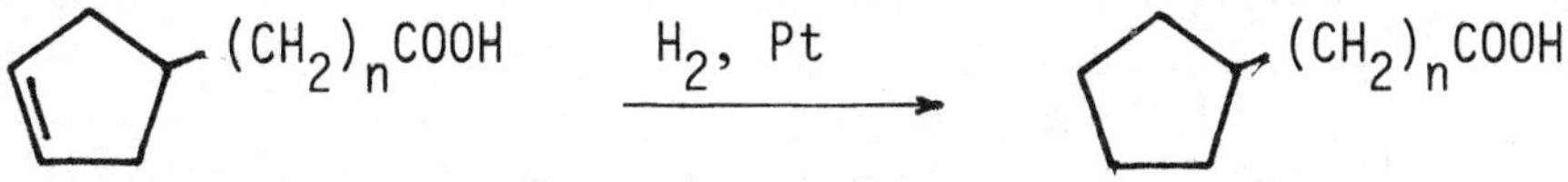

$(CH_2)_nCOOH$ —H_2, Pt→ $(CH_2)_nCOOH$

chaulmoogric acid, n = 12 → dihydrochaulmoogric acid

hydnocarpic acid, n = 10 → dihydrohydnocarpic acid

Scheme 2

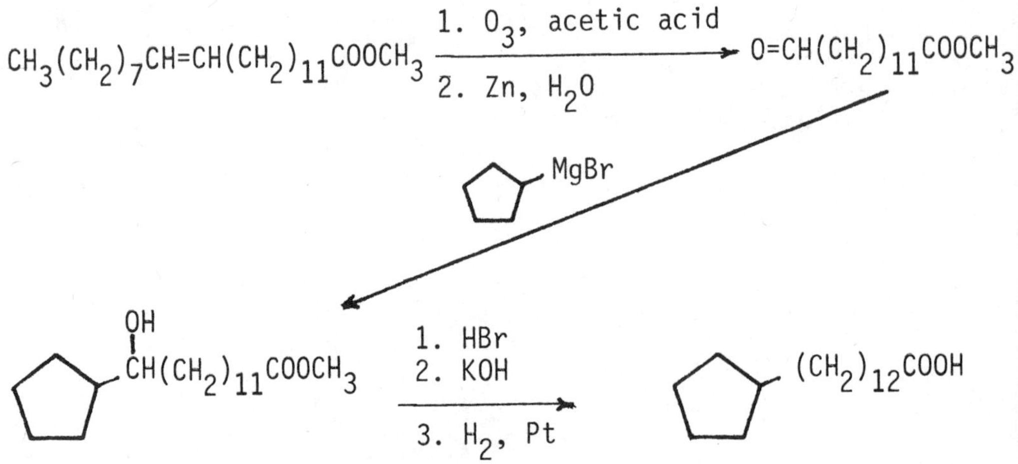

$$CH_3(CH_2)_7CH{=}CH(CH_2)_{11}COOCH_3 \xrightarrow[\text{2. Zn, } H_2O]{\text{1. } O_3\text{, acetic acid}} O{=}CH(CH_2)_{11}COOCH_3$$

$$\xrightarrow{C_5H_9MgBr} C_5H_9\text{–}CH(OH)(CH_2)_{11}COOCH_3 \xrightarrow[\text{3. } H_2\text{, Pt}]{\text{1. HBr; 2. KOH}} C_5H_9\text{–}(CH_2)_{12}COOH$$

Scheme 3

Variations on this type of synthesis were studied in over twenty papers and the resulting compounds were tested against *B. leprae*. Some clinical activity was found in ethyl dihydrochaulmoogric acid,[23] but the project did not result in clinically useful compounds. The chaulmoogric acid work was excellent chemically, however, and was always cited in the national honors Adams started to receive in 1927.

In addition to his seventy-three research papers from Illinois during 1916–26, Adams was responsible for the publication of six volumes of *Organic Syntheses* and the four bulletins on *Organic Chemical Reagents* during this period. (His Ph.D.'s from this time are discussed in a later section.)

Adams's record of significant scientific publications was due not only to his intellectual capacity but to his extraordinary talent for attracting and training students. As he said repeatedly throughout his career, he took great satisfaction in developing the ability of his collaborators to become productive, independent research workers, and the success of his former students was as important to him as his own success. Indeed, Adams measured his own achievement and that of his department in large part by the progress of former students. In spite of the large number of his research workers, he never regarded any of them as a "pair of hands," and he never sacrificed their welfare to promote his personal scientific advantage. This attitude was by no means universal among university professors, although it was general at Illinois. Adams's interest in his students as individuals was obvious to them and was one of the reasons for the respect and affection they felt for him.

He visited his research students frequently in their laboratories, sometimes in the daytime, sometimes in the evening, his usual greeting being, in the rather high-pitched Yankee twang that he never lost even after fifty years in Urbana, "Waall, what's new?" If little was new and if the research worker showed no inclination to talk, he passed on to the next worker. If there were new results or problems to discuss, he would settle down for a conversation, in which he usually made useful suggestions and sometimes proposed a look at some quite unrelated problem. Adams seldom forgot these off-the-cuff ideas and would usually inquire after a period of several weeks, "Did you try such and such?"

After the chemical discussion had run its course, Adams, unless he was extremely pressed for time, enjoyed discussing other topics—politics, current events, business conditions, sports, and other chemists (outside of Illinois), a subject on which his dicta were frank and sometimes ruthless. Adams's comments were always shrewd and usually pungently expressed, and his listeners realized that they were hearing a broadly informed and gifted mind at work.

Adams's dealings with his research group were always informal and never authoritarian in tone. His students in the 1920s knew him as "Rajah" or "The Chief," and the latter nickname became universal for the remainder of his career.[24] He enjoyed kidding and repartee and never discouraged a good argument on chemical or other topics. Nevertheless, he was not a backslapper and he never sought easy popularity; he was unmistakably "The Chief," and on no occasion was the narrow line between camaraderie and undue familiarity crossed. Adams expected, and got, hard and imaginative work from his students, his own intense interest in the research problems generating a similar interest on the part of his collaborators.

Adams developed a highly successful scheme for directing research. He started the student on a problem that promised reasonable success and thus built up the student's confidence. This usually avoided a long period of work without positive results, a particularly discouraging situation for a beginner in research. He then wrote up the first work for publication and assigned the student a more difficult and uncertain problem or sometimes several problems. He was criticized, as he said, for doing so much research on subjects such as hindered rotation in biphenyls and related compounds. Adams's view was that these problems gave the students excellent training in synthesis and in physical measurements, and he regarded this training as indispensable. Some of his natural-product research was very difficult experimentally, and he usually gave this work to postdoctorates, when he had them, and to experienced graduate students.

Adams wrote up research papers with extraordinary speed. When he felt that a piece of research had been carefully and thoroughly done experimentally, he asked the research worker for the written experimental section and sometimes a discussion section. So clear was his knowledge of the details and the key points of even complicated problems involving several workers, and so penetrating was his analytic ability, that he could block out a manuscript requiring only slight revision in a matter of hours. Adams wrote clearly, concisely, and in an admirable style for scientific papers. He had the "instinct for the jugular," getting to the heart of a problem and ignoring irrelevancies. Watching him write a paper was not only highly educational but a source of amazement to his collaborators.

A story, which is part of the Illinois legend, occurs in many variants, but the fact that it is so widespread is in itself significant. According to one version, Adams was roaming through the laboratories at night, found a graduate student poker game in progress, and asked to be dealt in. He played until he had won all the money on the table, then said, "Well, boys, this will teach you not to play poker in the laboratory," and departed. In the 1920s Adams and a few other faculty members played poker fairly regularly with a group of graduate students at the Gamma Alpha house; Adams was unconcerned that the graduate dean, who disapproved, lived next door and could see the party if he wished.[25] In his later career Adams's many trips limited his poker playing to consulting visits or to special occasions in Urbana.

In spite of his deep interest in his students and his success in bringing out their potentialities, Adams recognized clear limits to what he could do for them, which he expressed repeatedly. As he wrote to one of his very successful students:[26]

> I quite agree ... that a teacher's getting excited about something new in the research underway by his student is often something that impresses the student and he adopts unconsciously a similar attitude. But I think it is impossible to teach creativity, or I prefer to call it teaching a person to have imagination. I am convinced, as I have been for many years, that creativity or imagination is an innate quality in the individual and that using an arbitrary standard one person may have fundamentally twenty percent creativity, another eighty percent, etc. What education does is to help the student to use his creativity or imagination to the fullest extent possible in his make-up. The man who does not have the opportunity for such education will often never use his potential creativity more than fractionally. It is possible to train very able chemists and it is not necessary to have one hundred percent innate creativity to be eminently successful. It is a quality that is rarely found to a very high degree in scientists; as a consequence, the number of importantly creative scientists is very small compared to the many others. I am not depreciating the effort by a professor to instill in his students as much creativity as possible, but I feel that it is a development of a natural characteristic.

Adams felt that students should major in a science only if they found it fun; there was so much to be learned before they could experience the real excitement and satisfaction of positive results in research that they must have a real love of the subject to carry them to that point. This meant that their undergraduate courses must foster that fascination.[27] One of Adams's Ph.D.'s wrote:[25]

> I went to Urbana in the fall of 1920 with the vague intention of eventually studying medicine. It was not long, however, before I became more attracted to

and transferred to chemistry. There was already the beginning of the air of excitement in the Chemistry Department which was to characterize it through all of the Adams years. It permeated the undergraduate as well as the graduate school. As an undergraduate, I could only sense the excitement with little feeling for its source. In retrospect, it is amply clear that much of it centered around Roger Adams.

Adams dealt with people with perceptiveness and kindness. In his first conversation with a new postdoctorate from Harvard, he outlined the research problem he wanted investigated and then talked about his graduate students: many of them had not traveled much and got along on almost no money but nevertheless were highly capable chemists. The intent of this kindly homily was clear to the postdoctorate—Adams did not want anybody to treat his collaborators with condescension, nor did he want the postdoctorate to get off to a bad start by antagonizing his co-workers. The postdoctorate did not tell Adams that, in spite of his Harvard background, he probably had more experience at being an impecunious student than almost anybody at Illinois.

A second example comes from a woman graduate student of Adams's who described how Adams was proudly handing out cigars when his only child, Lucile, was born in 1927. One of the graduate students remarked, "What are you going to do for the women? They don't smoke cigars." That afternoon each woman graduate student found a box of candy on her laboratory bench.[28]

Many similar accounts illustrate the magnetism of Adams's personality and his insight in personal relationships. This understanding almost never deserted him, in spite of the pressure of his many activities and responsibilities. It was a rare student indeed who did not regard the association with Adams as a high point in his career. The letters presented to Adams on his retirement in 1954 speak eloquently on this point.[29]

LITERATURE CITED

1. O. Kamm, RA, and E. H. Volwiler, U.S. Patent, 1, 358, 751 (1920), *Chem. Abstr.*, *15*, 412 (1921); O. Kamm, *J. Am. Chem. Soc.*, *42*, 1030 (1920); RA and Volwiler, U.S. Patent, 1,676,470 (1928), *Chem. Abstr.*, *22*, 3265 (1928); E. B. Vliet, *J. Am. Chem. Soc.*, *46*, 1305 (1924). On other local anesthetics, W. R. Burnett, RA, et al., ibid., *59*, 2248 (1937).
2. RAA, 4, 1917–1918; contains the voluminous Abbott correspondence.
3. E. H. Volwiler, private communication.
4. Herman Kagan, *The Long White Line*, Random House, New York, 1963; a history of Abbott Laboratories.
5. RA, *J. Am. Chem. Soc.*, *41*, 247 (1919).
6. RA and H. Gilman, ibid., *37*, 2716 (1915); RA and L. F. Weeks, ibid., *38*, 2514 (1916).
7. RA and L. H. Ulich, ibid, *42*, 599 (1920); RA, W. V. Wirth, and H. E. French, ibid., *40*, 424 (1918).

8. RA and E. H. Volwiler, ibid., *40*, 1732 (1918); H. E. French and RA, ibid., *43*, 651 (1921).
9. Examples: RA and J. R. Johnson, ibid., *43*, 2255 (1921); RA and A. J. Quick, ibid., *44*, 805 (1922).
10. RA, M. F. Fogler, and C. W. Kreger, ibid., *44*, 1126 (1922).
11. RA and C. F. Rassweiler, ibid., *46*, 2758 (1924).
12. RA and I. Levine, ibid., *45*, 2373 (1923); RA and Edna Montgomery, ibid., *46*, 1518 (1924).
13. RA and V. Voorhees, ibid., *44*, 1397 (1922); preliminary account.
14. "A Sketch of Research Achievements," by RA, pp. 3–4, written about 1954; published accounts, J. H. Wolfenden, *J. Chem. Educ.*, *44*, 299 (1967); N. J. Leonard, *J. Am. Chem. Soc.*, *91*, a (1969); E. J. Corey, Welch Foundation, op. cit. p. 212. RA had told Wolfenden the story in 1949.
15. RA and W. H. Carothers, *J. Am. Chem. Soc.*, *47*, 1047 (1925); RA and B. S. Garvey, ibid., *48*, 477 (1926).
16. The preparation of the catalyst and the apparatus were described in detail by RA, V. Voorhees, and R. S. Shriner, *Org. Syn. Coll.*, Vol. 1, Wiley, New York, 1932, p. 463; RA and V. Voorhees, ibid., p. 61.
17. Morris Freifelder, *Practical Catalytic Hydrogenation*, Wiley-Interscience, New York, 1971, p. 10.
18. P. N. Rylander, *Catalytic Hydrogenation Over Platinum Metals*, Academic Press, New York, 1967.
19. R. A. Jacobson and RA, *J. Am. Chem. Soc.*, *46*, 1312 (1924).
20. Id., ibid., *47*, 283 (1925).
21. R. L. Shriner and RA, ibid., *47*, 2727 (1925); this contains references to the clinical use of these compounds.
22. C. R. Noller and RA, ibid., *48*, 1074, 1080 (1926).
23. G. S. Hiers and RA, ibid., *48*, 1089 (1926).
24. Edwin W. Shand to DST, April 24, 1978; Shand was an undergraduate at Illinois 1919–23 and in 1922 lived in a house with three of Adams's graduate students (McElvain, Dreger, and Calvery).
25. W. H. Lycan to DST, June 12, 1979; Lycan was an Adams Ph.D. in 1929.
26. RA to T. L. Cairns, September 20, 1967; RAA, 33, Du Pont.
27. Interview with RA, taped by C. O. Guss, January 20, 1961.
28. Beulah D. Westerman to Lucile Adams Brink, February 28, 1978.
29. The account of Adams's relations with his students by N. J. Leonard, *J. Am. Chem. Soc.*, *91*, a–d (1969), can scarcely be improved.

Academic Progress

The Illinois Department under Noyes and Adams

In 1926 William A. Noyes (born in 1857) retired as director of the chemical laboratories at Illinois, and Roger Adams was the unanimous choice of the division heads in chemistry to succeed him.[1] Adams never used the title "director" but was known thenceforth as "head of the chemistry department." Noyes's career was distinguished;[2] he published over ninety research papers from 1908 to 1936, trained more than a score of Ph.D.'s, served as editor of three American Chemical Society publications, traveled widely and wrote extensively in support of better international relations, and built up an outstanding department. Noyes's own research dealt with the chemistry of camphor derivatives, rearrangement reactions, and experiments on nitrogen compounds designed to test the dualistic theory of valence, which Noyes clung to long after G. N. Lewis had proposed the shared electron theory of valence. He wrote a series of widely used textbooks, his organic text going through four editions; it was notable for its broad view of the field and its use of physical chemistry. Noyes's textbooks were financially profitable; he did little or no industrial consulting. Adams, on the other hand, wrote only one text but consulted regularly for several firms.

It is a mistake to conclude that the large proportion of Illinois Ph.D.'s going into industrial research was due primarily to Adams, or that it represented a reversal of Noyes's policy. Before Adams was hired at Illinois, Noyes foresaw that industry would require many Ph.D.'s, as shown in the 1916 press release for the dedication of the new addition to the chemical laboratory.

The Noyes curriculum for chemistry majors in 1916 called for a year of inorganic chemistry and qualitative analysis, a year of quantitative analysis, a year of organic synthesis combined with quantitative and qualitative organic analysis, and a semester of physical chemistry, as well as electives chosen from electrochemistry, industrial chemistry, physiological and food chemistry, and bacteriology. Seniors were required to take research, history of chemistry, journal meeting (undoubtedly adapted by W. A. Noyes from Ira Remsen's journal meeting at Johns Hopkins), and chemical technology. Mathematics through calculus, a year of physics, French or German, and "rhetoric" (English composition) were also required.[3]

Graduate (advanced) courses were offered in electrochemistry, physical, inorganic, analytical, water, physiological, and animal chemistry, fuels, and metallurgy. In organic, an advanced course covered compounds of carbon, hydrogen, and oxygen, open-chain and cyclic; available also were quantitative organic analysis of proteins, alkaloids, volatile oils, and other constituents of plant and animal tissues, and a seminar in special topics in organic chemistry. Noyes and Derick supervised research in organic chemistry. In addition, students were kept abreast of recent research at Illinois and elsewhere by speakers at the local section of the American Chemical Society, by the Chemical Research Club, which met monthly for dinner and discussion of current research at Illinois, by the Chemical Journal Meeting, convening weekly, and by the Chemical Club, meeting monthly for social purposes and for lectures by "prominent speakers."

The honorary chemical fraternity Phi Lambda Upsilon (founded at Illinois in 1899) sponsored visiting research speakers. Alpha Chi Sigma, the national chemical fraternity, had a chapter at Illinois. Clearly Noyes planned carefully for the intellectual and social side of chemistry in 1916, but one suspects that many mugs of beer foamed outside of Noyes's carefully organized clubs.

E. H. Volwiler, a graduate student in 1916, recalls that the books used were W. A. Noyes's text in organic chemistry, with J. B. Cohen's treatise for advanced topics, Alexander Smith's general chemistry, and Nernst's book for physical chemistry; German texts were not much used, although a reading knowledge of French and German was required of Ph.D. candidates. Oliver Kamm, following Derick, was just getting organic qualitative analysis underway. Volwiler found that few of the graduate students were Illinois graduates but came rather from small colleges in the Midwest. Students in physiological chemistry and industrial chemistry did not have much contact with the organic graduate students during Volwiler's time.

The curriculum for chemistry majors in 1927 was not greatly changed, at least not in titles.[4] Physical chemistry was now a full year course; Miss Sparks, the chemistry librarian, taught chemical literature, and the chemical journal meeting had now disappeared for seniors, apparently being absorbed into the graduate seminars. A number of service courses were given, such as elementary organic chemistry for students in agriculture, and graduate students chose from a set of advanced courses. Adams taught elementary organic chemistry with a second semester including geometric isomerism, tautomerism, and mechanisms of important reactions, and J. R. Johnson ran the organic laboratory (one semester). Marvel lectured on special topics, and Adams supervised the organic seminar. Twenty-two faculty members offered research in twelve fields in 1927.

It is not clear from the course list of 1927 which was the organic qualitative analysis course. This course in identifying unknown organic compounds by their properties of solubility, behavior toward functional group reagents, and preparation of solid derivatives was a very effective American invention for the teaching of organic chemistry and was developed first at MIT by A. A. Noyes and S. P. Mulliken.[5] C. G. Derick initiated the Illinois course,[4] for which Oliver Kamm wrote a laboratory book[5] and in which Marvel played a key role. Thirteen years later Shriner and Fuson[6] wrote their classic text on the subject.

Adams treated Noyes, his retired predecessor, with great consideration and made him feel that he was still a valued colleague. Noyes retained space for experimental work along with a research assistant, usually a graduate student, though he did not take doctoral students. He continued his church activities, his strong interest in international affairs, and his speculations on atomic and molecular structure. He remained until his death, in 1941, a figure from the heroic age of organic chemistry in this country, who had worked with Remsen and Baeyer in the 1880s and was internationally respected as a scientist and as a man.

Adams's assumption of the headship of the department was not marked by any noticeable discontinuity from the Noyes regime, particularly since Adams had been directing a large research group since his return from Washington late in 1918. His standards of performance for himself and his students were as high as Noyes's, and he worked as hard. Nevertheless, there were changes after 1926, the most obvious being in size and in atmosphere. The marked increase in industrial research, the growth in college and university undergraduate enrollments, and the success of Illinois Ph.D.'s in both areas caused a steady growth in graduate enrollment and in size of staff, and Adams's enthusiasm charged the department with the desire to achieve.

Adams was head of the organic division from 1918 to 1926, and his place was taken by Carl S. (Speed) Marvel, a graduate of Illinois Wesleyan, with his Ph.D. from Noyes in 1916 and with extensive experience in preps.[7] Marvel resembled Adams in many ways: in candor, informality, originality, ability to inspire hard work and devotion in his students, and encyclopedic knowledge of organic chemistry. His productive research career covered more than six decades, one of the longest in American chemistry, and his work on polymerization was the first major project in the field in an American university. Among his other researches were studies on triarylmethyl free radicals and on acetylene chemistry. Marvel utilized physical techniques in his work and enlisted collaboration of physical chemists to a greater degree than did Adams.

Ralph L. Shriner, Adams's Ph.D. student, worked on natural products and was a staff member from 1927 to 1941. Reynold C. Fuson, a Minnesota Ph.D. and a National Research Council Fellow with Kohler, played a key role in the organic group from 1927 to 1963, his personality and research interests complementing those of his colleagues. An inveterate traveler, an accomplished linguist, knowledgeable in music as well as in chemistry, he excelled in teaching and inspiring students, both undergraduate and graduate. His research dealt with conjugated systems and steric hindrance.

Henry Gilman, who had a long, outstanding career at Iowa State, and Charles D. Hurd, a leading chemist at Northwestern for many years, were members of the Illinois organic division for short periods around 1920, as young Ph.D.'s. In the biochemistry division William C. Rose, who served from 1922 to 1954, became a distinguished scientist and a close personal friend of Adams; he was acting head of chemistry while Adams was in Washington during World War II. Herbert E. Carter, Adams's successor as head, was biochemist from the 1930s to 1967, and Vincent du Vigneaud, later a Nobel Prize winner, was at Illinois in biochemistry from 1929 to 1932.

Some of the other colleagues of Noyes and Adams, with their years of service at Illinois, were R. C. Tolman (1916–18) and Worth H. Rodebush (1921–55), a student of G. N. Lewis, both in physical chemistry. In inorganic chemistry B. Smith Hopkins (1912–41) and John C. Bailar, Jr. (1928–74) were leaders, and analytical chemistry included G. L. Clark (1927–60) and G. Frederick Smith (1921–57).[7]

Adams did not attempt to control the activities of the divisions beyond his authority over the allocation of the whole departmental budget. However, he did scrutinize all appointments, particularly tenure positions, with great care, and he planned at least two people of stature in each division so that there would be continuity of teaching and research if one left Illinois.[8]

Adams's statement of 1927[4] said that training in chemistry was necessary for students in engineering, ceramics, agriculture, home economics, and premedical and predental courses. Students with the B.S. in chemistry or in chemical engineering were equipped for industrial work or for advanced studies in chemistry. He continued:

> Chemistry is a subject undergoing rapid changes and every progressive university is contributing to development of the science. The staff, besides teaching the various courses, spends much time in research. There is no doubt a close connection between efficient teaching and productive research.
>
> Students, to learn methods of investigation, must continue their studies beyond the Bachelor's degree. A very important part of the best efforts of the department is spent in conducting this graduate work for students who are candidates for the degrees of Master of Arts or Master of Science, and Doctor of Philosophy. Though several years is often required before a student is able to do independent research, those who are so trained are in great demand, not only by universities to teach and to aid in training others in research, but also by the industries who need men for developing new processes and improving existing ones.
>
> The teaching staff of the department for 1926–1927 includes six professors, seven assistant professors, six associates, five instructors, forty-two half-time assistants, and eighteen quarter-time assistants. There are also ten fellows, six scholars, and nine research assistants.

Illinois University Archives

Reynold C. Fuson (1895–1979).

Illinois University Archives

William C. Rose (1887–   )

The total number of students registered in the courses of chemistry for the first semester of 1926–1927 is two thousand eight hundred and sixty-seven. The number of graduate students for the same period is one hundred and twelve.

The growth and research output of the Illinois organic division are shown in Table I.[9] These figures represent very nearly the complete output of the Illinois organic group for these years and also show the position of the Illinois department in total number of organic papers, compared to all other laboratories, university or industrial. The Illinois papers as a group compared well in quality and originality with work being done elsewhere in this country or overseas.

Table I. Number of Papers in Organic Chemistry from Illinois Published in *J. Am. Chem. Soc.,* 1914–1939

Year (Inclusive)	Illinois Papers	Relative Position
1914–18	17	Seventh
1919–23	71	First
1924–28	94	First
1929–33	150	First
1934–38	160	First
1939 only	66	First

High esprit de corps and relaxed friendliness, with contacts encouraged among all staff and students regardless of their research directors, characterized the laboratory. The informality and zest were particularly striking to one coming to Illinois from the compartmented reserve of the Harvard department. The cultural shock must have been severe for Wallace Carothers, who went in the reverse direction from Adams's Illinois to Richards's Harvard department in 1926; Kohler's letter in Chapter 4 implies this. Intense interest of the staff in organic chemistry, which naturally communicated itself to the students, added to this informal atmosphere. Adams obtained funds for modern equipment and instrumentation to facilitate research by every means possible, and the preps project produced a supply of organic chemicals as starting materials unrivaled in any other laboratory.

The spirit in the Illinois research groups was caught by W. H. Lycan:[10]

> [The organic faculty] ... knew not only their own students but those of each other and were deeply interested in all of them. Even by modern standards it was a large organic section but with the help of a very active organic symposium and in the complete absence of academic jealousy, every student felt free and frequently did consult any or all of the members of the quadrumvirate. Roger was the key to building this system of cooperation. He never hogged the most promising students, those with the best undergraduate credentials. On the contrary, he consciously steered many of Illinois' more illustrious students to his colleagues and was as happy for their successes as he was for those of his own students.
>
> The same spirit was carried over into the management of the department after he took over in 1926. It was amazing how closely he followed the work of students in all segments of the discipline and how well he kept track of them after graduation. Thus it seems to me that it is to him and, in only slightly lesser degree, to Speed Marvel that is due the fraternal spirit that came to characterize Illinois chemists. The popularity of the Illinois gathering and of Roger's annual report at the national meeting of the Chemical Society attest to the existence of a true fraternity.

What Adams accomplished by 1930 represented a significant advance in American university education, and the Illinois department became a model for other graduate departments of chemistry, not only in the state universities but in others as well. The only other American university at that time to develop graduate work in chemistry on a comparable scale was G. N. Lewis's department at Berkeley, and in some respects Adams's accomplishments exceeded Lewis's. Adams added a new dimension to that uniquely American institution, the land grant college.

Adams and his colleagues gave most careful consideration to the details of the graduate training program. The ideas summarized in the following paragraphs were expressed by Adams in other places, and this general scheme of graduate education was in operation for many years before it was explicitly formulated in 1950."[11] Indeed, the high point may have been reached in the 1930s.

The Illinois system employed group research laboratories accommodating four to ten graduate students, chosen carefully from different undergraduate schools and with varying degrees of experience. New students thus learned techniques and ideas from their more experienced colleagues, and discussion and cooperation flourished, generating an atmosphere of contagious enthusiasm for research. Students with different undergraduate training strengthened the learning process. The give and take in such a laboratory taught the students to get along with other people of differing backgrounds and personalities. The faculty research supervisor, using a subtle or if necessary a blunt approach, influenced this important development of personality, and organizations like Alpha Chi Sigma had an important role to play. The faculty promoted high morale in the graduate student group by sincere interest in the student's progress before and after the completion of the Ph.D.; as a result, the students felt that they were heirs of an outstanding tradition, which they must support by good work in their later careers. Students were expected to write and speak clearly, concisely, and forcefully and to learn about the operation of the chemical industry.

Faculty members were expected to be actively interested in chemistry, to maintain a position as productive scholars, and to set an example of unselfish cooperation with each other and with all graduate students regardless of the students' official research directors. "Students are quick to detect friction within a faculty." The departmental head's task was to keep the faculty members, new and old, working for the good of the department, not for their individual advantage, and to set a good example of scholarly productivity himself. In general he should appoint new staff from other institutions. All faculty members should play a full role in professional activities as models for the students, help to place them in appropriate positions, and recruit good graduate students from other institutions. Thus the requirements for a faculty position at Illinois were many. Adams and his colleagues were excellent judges of talent who made few mistakes in their appointments to faculty positions.

The influence of the Illinois department on other universities was great, partly by example and partly because of the excellent records of Illinois doctorates in positions of teaching and research throughout the country. Many Illinois graduates taught in liberal arts colleges. Organic chemists from numerous universities mingled on the boards of *Organic Syntheses* and later of *Organic Reactions*, and Adams, Marvel, and Fuson had close friends in many departments. Probably most important was the exchange of graduate students with other universities, particularly in the Midwest; Illinois graduates normally went elsewhere for Ph.D.'s and the Illinois graduate students came from other schools. Particularly close relations obtained between Illinois and Minnesota, Wisconsin, and Nebraska.

The essential features of the Illinois system were transplantable and were used in other universities with marked success. They were also applicable to industrial research laboratories, with obvious modification in detail, and many successful laboratories followed the general Illinois philosophy, regardless of the presence or absence of Illinois men on the staff.

The social and economic bases for the growth of graduate work in chemistry at Illinois and at other universities deserve detailed consideration. Because it is more logical to examine together all of Adams's Ph.D.'s from 1918 to 1958, the extended discussion is left for a later section.

The Great Depression brought serious administrative and human problems to the chemistry department, which Adams managed as well as he could, spreading the available funds as much as possible. Academic positions virtually disappeared for several years, except for strictly temporary ones. Government laboratories were not adding chemists, and the industrial laboratories offered the best opportunity for employment. No one who was professionally active as a chemist during the 1930s will forget the job stringency. Adams had no bias against academic positions, but as a pragmatist he advised students to take positions where they were available, which was mainly in industry.

The Depression affected Adams personally, as it did every other American. He started buying stocks in the 1920s, and there are some references in letters to Carothers about the stock market before the 1929 crash.[12] Adams suffered very serious losses in the stock market following 1929;[13] however, he recovered his financial equilibrium and continued to be a canny and successful investor for the rest of his life. A rather conservative Republican, as befitted his New England and family background, Adams strongly opposed Franklin D. Roosevelt and most of the New Deal measures. The chemistry department benefited directly by some of these, such as the work-study program that allowed some students to be paid for services as laboratory helpers. "Franklin and Eleanor" jokes, a phenomenon of the time among Roosevelt's opponents, were endemic in the laboratory.

Adams wrote later:[14]

> Another exciting, or perhaps it should be called discouraging, period for chemists was from 1931 to 1934. The country was in the throes of a deep depression, and we at Illinois felt most fortunate to place during those years about half our current Ph.D.'s in chemical jobs with salaries ranging from $1,500 to an occasional $2,000. Many Ph.D.'s who were not placed stayed on at the University, taking half-time teaching assistant positions at $600 a year; others took outside jobs—I remember one served as a gas station attendant, another as a clerk in a Sears and Roebuck store. But before June of 1935, all had been placed in chemical positions; and beginning with the inflation which was deliberately initiated by the government to help relieve the situation, the salaries of chemists started to rise and except for the second World War period when they were frozen by the government have been increasing ever since. And so has inflation.

Adams's correspondence with the Illinois administration indicates his watchful attention to departmental affairs and his attempts to ameliorate the effects of the Depression.[15] In 1931 he attempted to resign as head, saying that he could accomplish more for the university by research and teaching than by administrative work; the dean and president, by giving Edna Evans, his secretary, a higher title and more authority, persuaded him to continue as head. It is probable that this move on Adams's part was sincere and was not merely a power play.

How the Illinois department and the faculty perceived and ordered the graduate program in organic chemistry has been based partly on Adams's own words and partly on recollected personal experience. A far more immediate and direct record of graduate student life, as it appeared to a participant at the time, is given in a remarkable series of letters written over the years 1931–36 by an Illinois graduate student, Sidney H. Babcock, Jr., to Arthur W. Ingersoll of Vanderbilt University.[16] Ingersoll, a Nebraska graduate, took his Ph.D. with Adams in 1922 and continued to do excellent research on resolution of amines and related problems; he was an outstanding teacher. Babcock, born in Oklahoma in 1909, received his bachelor's and master's degree from Vanderbilt in 1930 and 1931, and Ingersoll, who kept in fairly close touch with the Illinois department, secured an assistantship at Illinois for him. Babcock wrote about fifteen letters to Ingersoll, describing in detail his career at Illinois, where he made a very good record. We have omitted the detailed chemical accounts of his research progress but have included other significant information from the letters.

The first letter of September 27, 1931, describes his initial impressions:

> My fears concerning Illinois have not materialized. So much contrary to my expectations has occurred that I am more than pleased with my situation, and am the more grateful to you for placing me here. All of the men on the staff from Dr. Adams on down have been most cordial. Despite the fact that there are 175 graduate students in the department, with about half of them in the organic division, I have not been made to feel I was in a mill with so many others, but that I was still an individual.
>
> Dr. Marvel, Dr. Shriner and Dr. Fuson have all advised waiting until the next semester to begin research. Dr. Adams however enrolled me in Chem 190 (the research number) with the understanding that if I were not in a position to do anything I could drop it without demerit. I think I shall follow the advice of the first three, for a summer's rust must be scraped off in the library, and also I need time to consolidate my position.
>
> My courses include a review course in Organic, a lecture course under Dr. Marvel; an elementary course in Physical Chem. (They have an antipathy toward Getman's text, the one used at Vanderbilt, so I have to repeat the lecture course without taking the lab work); and a course in "Conduction of Electricity thru Gases"; as part of my Second Minor in Physics. The total is 3 units. This part of my work should not be overly difficult.
>
> My teaching schedule is of 12 hours a week: 6 hours in the premed and home ec course laboratory, 2 hours taking roll during the lectures (by Dr.

Shriner) of the same course, 1 hour in charge of a quiz section in the same course, and 3 hours in the lab of the ag course under Dr. Fuson. The demands upon my time for this part of the work will be learned later as this week just begins regular work . . .

The liberality of the department with its apparatus, materials and general supplies is amazing—at least to one who for five years has been under Tom's regime. ["Tom" was evidently the stockroom custodian at Vanderbilt.]

His next letter of February 9, 1932, recounted a successful first semester, with A's received in his three chemistry courses and French and German exams passed:

My research was really negligible. Indeed, there was some justification for the remark of my office-mate, who said, when he learned of my A in research, "I don't see what he based it on!"

. . . As for teaching, I had no difficulty last semester, having the premeds, home ecs and ags. This coming semester, however, I have the chem majors in 37 [an organic laboratory course].

. . . I fear the instructor is to learn more than the students if the thing is to be successfully managed.

Incidentally, the Dean has given me full credit for the year at Vanderbilt. The work in physics completed a second minor. My room is the same used by Dr. Johnson of Cornell when he studied here.

His next letter of September 16, 1932, came after a year of good course work, passing his qualifying exams, a summer spent working in preps, and an unexplained and temporary disagreement with Marvel:

The summer of preps and study has ended, and the year's work is about to begin. Dr. Marvel, if he were growling about any particular person, has been thoroughly mollified by a fine and extremely successful fishing trip, so that he is in excellent humor. For the rest, the paycheck came within the following week after preps had ended.

Qualifyings were more or less successfully passed with A's in Organic, Inorganic, and Physical, and with a B in Analytical Chemistry. Over fifty took the examinations this time, several flunking out and two making four A's. I should have liked to have done this latter, but three weeks were hardly enough to overcome the five years absence from analytical phenomena.

I suppose you know of the new text "The Systematic Identification of Organic Compounds," by Shriner and Fuson, since your contribution is acknowledged in the preface. It will be another book to study for prelims, which will be around only too soon.

In his letter of October 16, 1932, Babcock describes his Ph.D. problem with Fuson and has interesting comments on preps, on his musical activities, which were not too common in the chemistry department, and other matters:

... There is no doubt in my mind concerning the wholesome influence of Summer Preps. Seldom, I suppose, has a reaction there had such a direct bearing on one's research; but the great variety of reactions run, the experience of handling diverse pieces of apparatus simultaneously, the economic necessity for turning out good products in fair yields make the 40¢ an hour but a small part of its value.

This semester I have sections of 32(Ags) and 34(Chem Majors); and Speed's Special Topics, Thermodynamics, Organic and Physical Seminars. Dr. Fuson has been very good to me. I shall be permitted to take prelims either in February or May. Dr. Fuson intimated that they were designed to see how much a man had assimilated rather than how much he had crammed down in heated pre exam special study, and advised me to adopt a study program now and wait until May, for the things I studied now for an exam so far away would be more likely worthwhile knowing.

Dr. Clark of the Analytical Division, Dr. Kubitz of the department of Philosophy, and Dr. Audrieth of the Analytical Division (lately returned from Germany) and I have formed a string quartet. We meet Monday evenings and have a great time together coursing thru the works of the masters—Brahms, Beethoven, Haydn, etc.

Room and board are but $30 a month this year as compared with $45 last year; this corresponds to a greater increase in salary than I was supposed to get, altho it actually stayed the same. Since last summer I have bought Lewis & Randalls' *Thermodynamics* and Stewart's 6 ed. of *Recent Advances*. I plan also to buy Bernthsen & Sudborough, Gatterman-Wieland (Eng. trans.—Fuson plans to use it in 37 next semester) and Wittig's book. This will be about all the budget will stand for a time.

Altogether there is every augury of a great year ahead: life is fuller and more pleasant than ever before. My relations with the other men are more pleasantly established and I feel more like I "belong."

Next summer he was working in preps again and had become acutely aware of the Depression (letter of June 21, 1933):

With the closing of preps this summer I shall have finished a hectic twelve months since last seeing you. Studying for qualifyings, taking them; the last semester of course work, preparing for prelims, taking them; thru it all the steady undercurrent of research, teaching, study, and music—all has passed so rapidly as to leave me now with a very dazed feeling—so that I can hardly get up the ambition to accomplish anything.

The let-down after passing prelims has been accentuated by a sort of world-weariness that fills me. Life could be so simple and sweet by the exercise of the smallest portion of our general intelligence. Yet the world is such a topsy-turvy, nonsensical, uncertain sort of proposition as hardly to be conducive to ambition. In a vague way I realize that now is the time for the able man to assert his ability; that out of our present shake up, those of us who put in the best licks will somehow achieve some stability. But to spend some two weeks, as I have just done, among the jobless chums of my childhood; to return and watch the mournful mien of half the Ph.D.'s of this year's crop—as yet unplaced, puts a bitter taste to whatever small successes I have thus far had. And so a good part of my time I spend looking for something solid on which to predicate my existence. Maybe it's better just to exist and be done with it.

A few days later he wrote again, partly about his preps work, but more movingly about the effect of the Depression on his home town, Holdenville, Oklahoma (letter of June 30, 1933):

Real prep weather is with us, which should be enough to describe the present state of physical comfort. Thus far I have checked two preps: diphenyl methane in 50% yields from benzene, benzyl chloride and 10 grams of amalgamated aluminum catalyst; and diphenyl sulfide in 80% yields from benzene aluminum chloride and sulfur monochloride.

Dr. Marvel says to tell his latest is 38 2-lb bass in a single day's catch. Believe it or not! ...

I suppose a person's first bump is the hardest: and the fact that I "suffer" this particular bump only vicariously makes complaining on my part unwarranted. As you have remarked before fortune has been more than kind to me. I had just come back from Holdenville when I wrote you. Father is an official of the local Red Cross organization, and a director of the Chamber of Commerce. He is therefore concerned with much of the relief work of the county, and I obtained a cursory understanding of the extent of the misery:

Two persons committed suicide while I was there. One a butcher unemployed, the other a widow with children she could no longer provide for.

A man was mowing our yard who ten years before had been a hard working, money making contractor. He lost out when the bottom dropped out in Amarillo. They moved to Holdenville because his wife owned a small house there. The furniture bespeaks former comfort. Too proud to accept charity, too honest to thieve, he mows yards, his wife takes in laundry.

Only one of my childhood acquaintances has found gainful employment. The rest twiddle their thumbs; get the malignant jitters from poison liquor and create the "young people's problem." The town was particularly hard hit by "jakeitis," paralysis resulting from a bad shipment of "Jamaican ginger."

The firm my uncle travelled for folded up when its bank went under. He is sitting at home with a wife and four sick children, penniless, making the problems seem more close at home to us.

True enough, these things are not civil war, and utter ruination. But it all seems so strangely unlike the order and sanity that prevailed in my childhood surroundings as to be very saddening. In time I shall become inured to it, and meanwhile I only hope, for my own part, that I can contribute something to increase the orderliness of society.

... One phase of my research is concluded, the results of which will appear in the July issue probably. However, I've not got much to crow over, for I overlooked a reference which appears to make my work unnecessary. Hence we'll be starting something new next year.

The all-important question of a job was concerning Babcock when he wrote after Thanksgiving Day, November 30, 1933:

From a selfish point of view, a definite upturn in chemical industry is the thing most to be thankful for. In an interview with Tanberg of the Du Pont experimental station, for example, I learned that Du Pont's had declared an extra dividend. At any rate the big companies are scrambling for men. Du Pont's have taken six, will take more; and Rohm and Haas two, of the eighteen graduating from here either in February or June. "Speed" thinks I shall get an offer from Eastman's to work at their Kingsport plant. In this connection will you write me by return mail, *please,* what you know about the opportunities there?

I had a long talk with Speed yesterday. (We have, by the way long since gotten over the things I mentioned to you that prep-day summer before last, to the extent that he selected me to be one of the two assistants in 38 (analysis) this year, for which I felt recompensed for all my worrying.) [Chemistry 38, referred to here, was the famous course in qualitative organic analysis.] The advice here is almost unanimous that industry for the next ten years will be far better than teaching. If however I decide to embark on a teaching career, they have indicated a willingness to back me for a National Research Council Fellowship. They say that if I am lucky I might get somewhere, but that otherwise the teaching world seems to be a closed corporation. At the same time Speed mentioned a half dozen men such as Conant and Kohler of Harvard, Gomberg, Reid, who are quitting their posts for one reason or another, leaving opportunities for the rest of us. Few men, however, could have had the spectacular rise that Fuson has enjoyed; yet their success drives others of us on.

I had always taken for granted that I was to enter the teaching profession. But of late industry has been painted in such rosy hues as to place me on the fence in the matter. Your advice would be greatly appreciated.

Incidentally, there are about a half dozen organic staff men graduating this year, and about an equal number of men from other divisions. I asked Speed if I could let you know about it, and he said, "By all means." So if you have someone ready to come, the time is about ripe.

Life proceeds in a very rounded and happy fashion. Teaching, research, string quartet; term paper on alkaloids, a graduate student who plays the piano and lately, a young lady! and time seems non-existent.

By May 19, 1934, the date of his next letter, Babcock had completed his dissertation and was applying for a postdoctorate at Harvard as well as a National Research Council Fellowship. Adams received two postdoctorates in February 1934 as a result of his offer of a professorship at Harvard, and Babcock won an appointment. "The Chief" is, of course, Adams in this letter of May 19, 1934:

Since Dr. Fuson paid you a visit after the Florida meeting, I suppose he told you concerning the things that have befallen me since last I wrote you. It makes for such good talking, though, that I'll write you all about it anyhow.

First, I want to thank you for sending me the reprints: they came the next day after I had wired you, and were promptly forwarded to Harvard. I waited and waited for news from them—but being pretty busy and having Speed assure me with a broad grin that I wouldn't lack for something to do kept me from getting anxious.

Finally, one day the Chief stopped me in the hall and asked me to come see him the next morning. He then told me that if I didn't hear from Harvard favorably he could offer me an eleven-month research assistantship for next year at a salary a good bit larger than either a National Research Fellowship or the awaited Harvard fellowship. He told me to wait a week or so before deciding. It took me only an hour.

The Chief is reentering the local anaesthetic field. He has a variety of problems that are of great interest as well as practical importance. I had a long talk with him yesterday morning concerning the work. I could readily see why he has become the head of the society; and I welcome the opportunity to be with him.

In addition to research on local anaesthetics, I will be expected to lecture in his absence and to look after his new research students. I thus have an opportunity to become accustomed to most of the phases of an academic chemist's life. Naturally, I look forward to a very happy year.

For the summer, I have been given a full-time instructorship to lecture in 33—the elementary course for home-ecs and premeds.

... Doing research for a Ph.D. is hardly a satisfactory procedure from the research point of view: The first year, course work; second year prelims; third year, thesis—together with a multiplicity of odd duties that continually interfere. After a year is ended and the work is summarized, it is always amazing how little has been accomplished. Yet most of the academic research that is done at all is done under these circumstances.

In the next letter of December 18, 1934, Babcock tells of his teaching for Adams while the latter was away as president of the American Chemical Society:

Next semester, in addition to lab work, I shall, at least according to present plans, lecture in stereo when the Chief is absent. As he has been away between 60 & 70% of the time this semester it appears that quite a bit of "pinch lecturing" will be necessary.

My lab partner, a Dr. Knauf, and I visited the laboratories of Abbott's & Eli Lilly's during the Thanksgiving holidays. They were the first commercial labs I had seen, so that I was greatly interested in the trip.

According to his letter of May 2, 1935, Shriner had asked Babcock to do some writing on the chapter on stereoisomerism for Gilman's *Organic Chemistry, An Advanced Treatise*. Because of Ingersoll's work on resolution, Babcock was experienced in this field, and he discussed some technical points with Ingersoll. Babcock's outline (omitted below) is followed closely in the printed version of the chapter.[17]

After various handling, the part of Gilman's book on Stereoisomerism has come into my hands for quite a bit of rewriting and extension. Among other things I am rewriting the part on Methods of Resolution according to suggestions embodied in a letter from you to Dr. Shriner last September. I was given carte blanche as to treatment of the material. For my own guidance, I have made the chapter on Resolutions part of a section entitled "Racemic Modifications.

... The world is treating me most excellently these days. I shall be here at Illinois on the same appointment next year.

On September 13, 1935, after describing a serious eye accident in preps to one of his Vanderbilt contemporaries, Babcock goes on:

One of the Du Pont–Illinois boys was back for a visit recently. He said Tanberg had told him they were building a new lab and intended to hire 40 more Ph.D.'s within two years—it is rumored they plan entering the pharmaceutical game by acquiring Lillys!

C. S. Hudson was thru this summer. The Bureau of Public Health wants 10 men with Post Ph.D. research experience to begin a lab in pure chemistry on vitamins, hormones, etc. The Chief & Speed talk it up as a great thing.

Ruzicka [from the Federal Technical Institute, Zurich] also paid us a visit this summer. The Chief gave us a dinner for him which will be long remembered.

At present I'm learning enough about chlorophyll chemistry to make the first seminar talk using Fischer's papers....

Whatever he wrote for Gilman, Babcock apparently satisfied Adams by his teaching, writing, and research, because he stayed two years as postdoctorate and published a paper.[18]

Babcock wrote the last letter in the series on February 24, 1936, from the College of Agriculture of the University of California at Davis, then a college of about a thousand students. He had taken an instructorship there on Adams's advice. Although Babcock was the first organic chemist at Davis and found nothing on hand for research equipment, he apparently liked the position, because he stayed until 1943, at which time he moved to Lederle Laboratories. He spent the rest of his career at Lederle and with the parent company, American Cyanamid, in responsible administrative positions.

Written with no thought of publication, but for the information of his respected teacher and friend, Ingersoll, Babcock's letters give a vivid and perceptive account of one graduate student's experience at Illinois, when Adams and his colleagues were guiding one of the country's best divisions in organic chemistry.

Literature Cited

1. W. C. Rose, private communication; President David Kinley to S. W. Parr, January 23, 1926, thanking him for "telling me that the Heads of the Divisions in the Chemistry

Department endorse Dr. Adams for the Directorship of the Laboratory," Archives, Dean's Office, Department and Subject, 15/1/1, 17, Chemistry, 9/25–12/25.
2. RA, *Biog. Mem. Nat. Acad. Scis., 27,* 179 (1952); biography of Noyes.
3. *Circular of Information of the Department of Chemistry, University of Illinois, 1916,* passim, Archive, Liberal Arts and Sciences, Dept. Histories, 15/5/0/1, 1.
4. *Special Circular of the Department of Chemistry, 1916–1927,* August 30, 1927, Archives, Ref. 3.
5. The first text on qualitative organic analysis seems to have been one by A. A. Noyes and S. P. Mulliken of MIT in 1897, reviewed by Ira Remsen in the *Am. Chem. J., 20,* 251 (1898). Noyes turned to physical chemistry and built up outstanding schools at MIT and later at Caltech. Mulliken spent a lifetime of patient scholarship on qualitative analysis, published in four large volumes, 1905–22. He listed over 10,000 organic compounds and many dyes. His pioneer publications were followed by H. T. Clarke in 1912 and by Oliver Kamm, *Qualitative Organic Analysis,* Wiley, New York, 1923, reprinted many times and issued as a second edition in 1932. In 1905, according to Kamm, qualitative organic analysis was taught in only two or three universities, in 1915 in 15 or 20, and during the 1920s and 1930s the course became a standard feature of most undergraduate curricula for students planning to concentrate in organic chemistry.
6. R. L. Shriner and R. C. Fuson, *The Systematic Identification of Organic Compounds,* Wiley, New York, first ed., 1935; many later editions.
7. This and information about other personnel is in *Centennial 1967: Department of Chemistry and Chemical Engineering, University of Illinois,* 1968. More detail is given by G. D. Beal, in Ref. 4, pp. 9–23.
8. W. C. Rose, private communication; taped interview by DST with C. S. Marvel, 1977.
9. D. S. and A. T. Tarbell, studies on the development of organic chemistry in the United States.
10. W. H. Lycan to DST, June 12, 1979.
11. RA, R. C. Fuson, and C. S. Marvel, *Ind. Eng. News, 28,* 2765 (1950).
12. RA to W. H. Carothers, RAA, 7, Carothers.
13. Private communication, Mrs. W. G. Palmer. Conant, p. 115, tells of a professor who had been saving capital for his old age from consulting. He told Conant early in the 1930s that he had thought he was well off but now had no money left. Although Conant gives no name, RA is almost certainly meant.
14. RAA, 44, Noller Symposium; lecture by RA at Carl Noller's retirement dinner, May 20, 1966.
15. Correspondence with Presidents Kinley and Willard, and Deans Babcock and McClure, Illinois Archives for each officer.
16. Vanderbilt University Library, Special Collections, A. W. Ingersoll Papers, Babcock Correspondence, folders 1–9, quoted with the kind permission of Sidney Babcock.
17. Gilman, 1st ed., Vol. 1, pp. 181–96. Babcock is not given an acknowledgment, but he evidently thought about and worked on the section with care.
18. S. H. Babcock and RA, *J. Am. Chem. Soc., 59,* 2260 (1937).

Adams's Students: Origins and Careers

The growth of the graduate program in chemistry at Illinois was in some degree a consequence of the economic and social conditions in the country. However, Adams's own activities influenced these conditions as far as chemistry was concerned so that he was in himself a social force. His group of 184 Ph.D.'s over the years 1918–58 and about 50 postdoctorates from 1934 to the 1960s gives a valuable sample for studying the origins and careers of American chemists. The results show, among other things, that graduate study in chemistry was an important pathway to upward social and economic mobility.

The figures on doctorates granted in all fields of chemistry during Adams's career are given in Table II; organic chemistry cannot be distinguished from other branches, except that Adams's students were all organic chemists.[1,2]

From 1920 through 1944 Adams trained 137 Ph.D.'s but by 1942 his war research activities in Washington were seriously curtailing his academic research program. The Illinois department in the same period trained 661 Ph.D.'s, Adams thus accounting for twenty-one percent of this total. From 1920 to 1929 Adams trained forty-two percent of all Illinois Ph.D.'s in the chemistry department. As his colleagues became better established in research, his fraction of the total Ph.D. crop fell, a fact partly due to his success in helping his younger colleagues develop their own research programs. The Illinois Ph.D. production of 661 from 1920 to 1944 was 7.7 percent of the national total of 8,627 in chemistry.

Table II. Doctorates Granted in All Fields of Chemistry, 1918–58

Years	Illinois[a]	Total U.S.
1918–19	19 (2)	—
1920–24	64 (30)	746
1925–29	71 (26)	1178
1930–34	121 (25)	1751
1935–39	165 (22)	2212
1940–44	240 (34)	2740
1945–49	218 (13)	2586
1950–54	303 (20)	5172
1955–59	260 (12)	5068

[a]Figures in parentheses represent Adams's own Ph.D.'s for these periods.

It is frequently said that strong emphasis on graduate work damages the quality of undergraduate teaching; this thesis was certainly wrong at Illinois. The effectiveness and appeal of the Illinois undergraduate teaching in chemistry is shown by the fact that over the 1920–61 period, 721 people with bachelor's degrees from Illinois got Ph.D.'s in chemistry, the largest number for any institution in the country; the next departments on the list are City College of New York with 561 and Berkeley with 519.[3]

The years 1920–60 showed a striking increase in the number and size of industrial employment opportunities for Ph.D. chemists. It is estimated that in 1920 there were 300 industrial research laboratories in the United States and that by 1940 the number had grown to 2,200. The chemical industry was believed to have 3,400 research people in 1927, and, in spite of the Depression, the number increased to 9,542 in 1938.[4] Another set of figures shows large increases in people employed in research between 1938 and 1962.[5] The same trend is shown by U.S. Census reports.[6]

Figures in Table III for the total number of bachelor's degrees awarded and faculty size in the country show the large increase from 1920 to 1950. It cannot be proved that a constant fraction of these graduates were in chemistry or took some chemistry courses, but that is a reasonable guess.

In 1926 when Adams became head at Illinois, 252 Ph.D.'s in chemistry were granted in the whole country;[1] Illinois gave 18, of whom 6 were Adams's students. In 1950 the annual figure for the whole country was about 1,000, and the Illinois contingent was 74.

Table III. Bachelor's Degrees Awarded and Faculty Size in American Colleges and Universities[7]

Year	Faculty[a] (thousand)	Under-graduates (thousand)	Degrees[a] (thousand)	Degrees in chemistry[b]
1920	43	462	47	—
1930	72	924	111	—
1936	88	1063	125	—
1940	132	1494	187	—
1944	134	1155	126	—
1948	196	2616	272	—
1950	210	2659	434	10,600
1953	—	2231	305	6,000
1958	345	2631	366	7,000

[a]The basis for the figures in the *Statistical Abstract* changes, so that not all the figures are exactly comparable, but the trend is clear.
[b]Earlier figures not available.

Figures in Table IV on membership in the American Chemical Society also show the increase in numbers of chemists with a bachelor's degree or higher.[8] The largest increase is after World War II.

Table IV. Membership in ACS

Year	Members	Year	Members
1910	5,081	1940	25,514
1920	15,582	1945	43,075
1926	14,704	1950	63,349
1930	18,206	1955	75,233
1935	17,541	1960	92,193

The growth in employment in industrial research meant a corresponding increase in undergraduate chemistry enrollments, which in turn meant more positions in teaching chemistry in colleges and universities. It is notable that a larger proportion of Adams's Ph.D.'s went into academic work in the early 1920s than later.[9] The work of Adams and his colleagues as industrial consultants helped in establishing close relationships between the Illinois department and industrial research laboratories, and the example was followed by other university departments.

Chemistry attracted ambitious students in part because it offered a path of upward social mobility to poor but capable students, those who could not afford the long and expensive medical education, for example. Statistics on the economic and class origins of graduate students in organic chemistry would be very difficult or impossible to obtain. However, it is probably true that only a handful of the Illinois graduate students came from upper-middle-class families. Some indication can be obtained from the undergraduate colleges of the 184 Adams Ph.D. students, 1918–58, of whom seven were women.

Thirteen of Adams's Ph.D.'s were University of Illinois undergraduates, of whom eight started graduate work before 1926 when Adams became head, and it became unofficial policy to encourage Illinois undergraduates to pursue graduate work in chemistry at some other university. The University of Nebraska was next with ten, reflecting the influence of Cliff S. Hamilton, an inspiring teacher and close friend of the Illinois organic faculty. Oberlin had six; Washington University at St. Louis, Penn State, and Oregon State had four each; while Miami of Ohio, Denver, Indiana, Amherst, Ohio State, Michigan, Tsing Hua (China), and Lehigh had three each. With two each were twenty-three schools: the University of Washington, Grinnell, Connecticut Wesleyan, Tarkio, North Central (Illinois), Chicago, Carleton, Mount Holyoke, Hope, Illinois Wesleyan, Western Reserve, University of California at Los Angeles, Montana State, Harvard, Dartmouth, Berkeley, City College of New York, Minnesota, North Carolina, James Millikin, Illinois State Normal, Cornell, and Missouri School of Mines. Seventy-two colleges or universities had one each: fifteen were in the Northeast, twenty-six in the Midwest, six in the Far West, fourteen in the South, four in Canada, two in China, three in India, and one each in Sweden and Austria. Of the sixty-one American schools with one student each, only fifteen were major universities with good graduate programs, and thirty-one were small liberal arts colleges, drawing students mainly from their own immediate area. The others had no or very small Ph.D. programs. Although students going to small local colleges are not necessarily financially poor, many of them are, as are many of the students going to state universities, city colleges, or private universities. There is thus good reason to believe that the majority of Adams's graduate students were from families with modest incomes.

As Table V shows, over half of Adams's students (counting University of Illinois graduates) came from midwestern institutions, as was natural. However, the other sections of the country were well represented.

Table V. Undergraduate Origins of Roger Adams's Ph.D.'s, 1918–58

Area	No. of Students	Percentage of Students	No. of Colleges
Northeast	37	20	24
Midwest	85	46	46
Far West	18	10	11
South	16	9	15
Illinois	13	7	1
Foreign	14	8	12
Total	183[a]		109[a]

[a]Unknown: one U.S.

Of Adams's fifty-three postdoctorates, academic origins of the Ph.D. of the forty-five Americans are known, and the remainder were from foreign universities. The Universities of Minnesota, Illinois, Maryland, and Harvard each furnished three, and the following furnished two each: Wisconsin, Ohio State, California at Los Angeles, Texas, Northwestern, Columbia, and Rome. Nineteen universities sent one each.

Of Adams's Ph.D.'s, fifty-nine (thirty-two percent) went into teaching upon graduation, forty-two of them before 1930. Of the forty-nine who ended up in teaching careers, thirty-four (sixty-nine percent) attained the rank of professor, and of these, fourteen additionally became department head or dean. The list includes several who made important contributions to organic chemistry and who were excellent teachers: A. W. Ingersoll (1922), Vanderbilt; J. R. Johnson (1922), professor at Cornell and one of the original editors of *Organic Syntheses* and *Organic Reactions*; S. M. McElvain (1923), Wisconsin; R. L. Shriner (1925), professor at Illinois and department head at Indiana and Iowa; C. R. Noller (1926), Stanford; E. B. Riegel (1935), professor at Northwestern and later director of research at G. D. Searle; N. Kornblum (1940), Purdue; and Jack Hine (1948), Georgia Tech and Ohio State. Several were distinguished in the biological sciences; W. D. Langley (1922), head of the Department of Biochemistry at University of Buffalo School of Medicine; A. J. Quick (1922), director of the Department of Biochemistry at Marquette School of Medicine; Wendell M. Stanley (1929), who became chairman of the department and head of the Virus Laboratory at Berkeley; B. R. Baker (1940), a leader in medicinal chemistry at Santa Barbara; W. D. Fraser (1941), head of the Department of Microbiology at Indiana; and John M. Stewart (1952), professor of biochemistry at Rockefeller University and at the University of Colorado Medical School. All of Adams's Chinese Ph.D.'s returned to China to teach in universities. Thus although the percentage of Adams's students going into academic work was not large, the *absolute* number was large, which gave Illinois a strong influence in the academic as well as in the industrial research world.

Adams had a strong orientation toward the chemical industry, although he encouraged his students who wanted academic careers. The effect of the Depression in limiting academic positions more than industrial positions is set forth earlier. It is not surprising that, upon graduation, 120 (sixty-five percent) of his students entered industry, in almost equal numbers for each of the four decades of his teaching career. Of these, 107 remained in industry, 13 left (primarily for teaching), and 12 came to industry from other positions, for a total of 119 in industrial careers. Of these, fifty (forty-two percent) attained a position of director of research or equivalent, and fourteen of these eventually became members of higher management.

E. H. Volwiler (1918) became chairman of the board of Abbott Laboratories. Others who attained executive positions were E. E. Dreger (1924), vice-president for research, Colgate Palmolive Peet; C. Rassweiler (1924), vice-chairman of the board, Johns Manville; W. H. Lycan (1929), vice-president for research and member of the executive committee, Johnson & Johnson; W. E. Hanford (1935), vice-president of research, M. W. Kellogg and at Olin; E. E. Gruber (1937), vice-president for research, General Tire & Rubber; D. J. Butterbaugh (1938), president of Micromedic Systems, Rohm & Haas; T. L. Cairns (1939), director of Central Research Department, Du Pont; R. O. Sauer (1941), corporate manager Europe, Universal Oil Products; R. B. Wearn (1941), vice-president for research and development, Colgate Palmolive International; and R. M. Joyce (1938), department research director, Du Pont.

Probably the outstanding technical achievements were made by Wallace H. Carothers (1924), who moved from Harvard to Du Pont to head its new fundamental research group; his spectacular success with polymers gave a large boost to industrial research and was the progenitor of the many developments based on condensation polymers. Other significant technical contributors included G. D. Graves (1923), who played a key role in the nylon development at Du Pont and later became general director of research divisions of its Textile Fibers Department; M. M. Burbaker (1927), also prominent in the nylon development and later assistant director of Du Pont's Chemical Department; J. F. Hyde (1928), who worked on the development of silicones at Dow Corning; E. B. Riegel (1935), professor at Northwestern and later director of research at G. D. Searle; R. C. Morris (1936), senior research associate at Shell Development; M. T. Leffler (1936), director of research liaison at Abbott Laboratories; and M. W. Miller (1940), group technical director, Photo Products Division, 3M.

Only five (three percent) of Adams's students went directly into government laboratories, and a total of sixteen (nine percent) eventually made their careers in them. W. R. Brode (1925), professor at Ohio State, later became scientific advisor to the secretary of state. G. R. Yohe (1929) headed the division of coal chemistry at the Illinois State Geological Survey, and W. I. Patterson (1934) became director of the division of agricultural research at the U.S. Department of Agriculture Laboratory. Allene Jeanes (1938) was research chemist at the Peoria Regional Laboratory of the USDA.

Of Adams's forty-five postdoctorates who made their careers in this country, nineteen were primarily in academic work, twenty-three in industry and two in government. Of Adams's Ph.D.'s and postdoctorates, including his first (undergraduate) research student, Henry Gilman, ten were elected to the National Academy of Sciences, and most of the postdoctorates held positions comparable in responsibility to those of his Ph.D.'s.

It is a safe conclusion that for the majority of Adams's students, graduate work in chemistry opened the door to better and more satisfying careers than their fathers had had. Based on personal knowledge of the careers of hundreds of Ph.D. chemists, this conclusion seems generally true. It certainly was true of Adams himself, and he is quoted as saying of the Ph.D.'s of the mid-1930s,[10] "These kids all come in rabid Democrats and as soon as they get out and get a job and earn a living they become Republicans." This is really a restatement of the social mobility idea.

There is an interesting contrast between the career paths of Adams's students as above and those trained by Ira Remsen at Johns Hopkins from 1878 to 1913.[11] Remsen's students came mainly from the Northeast and the South; they went predominantly into college or university teaching and few went into industry. This contrast is a reflection of the difference in personal outlook of Adams and Remsen and also of the less highly developed state of the chemical industry in this country in Remsen's time. The feature of upward social mobility through graduate training in chemistry was common to both groups.

Literature Cited

1. Figures on doctorates taken from L. R. Harmon and H. Soldz, *Doctorate Production in United States Universities 1920–1962,* National Academy of Sciences—National Research Council, Washington, D.C., 1963, pp. 10, 20; figures for Adams's

Illinois University Archives

Adams and E. F. Rogers, ca. 1930.

Illinois University Archives

Wallace H. Carothers (1896–1937).

students from an unpublished list. The figures for Illinois given by Harmon and Soldz are slightly smaller than those of Ref. 2, because the latter includes doctorates in biochemistry and industrial chemistry in addition to chemistry.

2. *Centennial 1967: Department of Chemistry and Chemical Engineering, University of Illinois, 1968.*
3. Ref. 1, Appendix 4, pp. 89, 91, and 98.
4. *Research—A National Resource,* 3 vols., National Resources Planning Board, Washington, D.C., 1938–41. Vol. 2: *Industrial Research,* R. Stevens, p. 37, and F. S. Cooper, p. 180.
5. K. A. Birr, in *Science and Society in the United States,* D. VanTassel and M. Hall, eds., Dorsey Press, Homewood, Ill., 1966, p. 69.
6. U.S. Bureau of the Census, *Statistical Abstract of the United States, 1964,* pp. 540, 544; *1960,* pp. 542, 538. Figures on scientific research become much more detailed after the National Science Foundation started compiling them in the 1950s.
7. *Statistical Abstract of the U.S., 1939,* p. 109; *1948,* p. 135; *1950,* pp. 123, 124, 128; *1951,* p. 124; *1954,* pp. 134, 139; *1959,* p. 128; *1960,* p. 129; *1961,* pp. 124, 127. Numbers have been rounded off.
8. H. Skolnik and K. M. Reese, eds., *A Century of Chemistry,* American Chemical Society, Washington, D.C., 1976, p. 456.
9. We are indebted to R. M. Joyce for his help in tracing Adams's Ph.D.'s and postdoctorates.
10. T. L. Cairns to E. J. Corey, September 28, 1976.
11. D. S. Tarbell, A. T. Tarbell, and R. M. Joyce, *Isis, 71,* 620 (1980).

Service and Research to 1942

Scientific Citizen, 1918–42

In the period between the world wars, Roger Adams became the most influential organic chemist in the country and was asked to serve chemistry and science in many capacities. He felt a strong sense of duty to the scientific community and to the nation to undertake public service activities that he considered important. Adams never confused the shadow with the substance, and therefore he never undertook activities that could be predicted to be merely busy work. Conversely, if he did accept responsibility, he gave it his concentrated attention and made a useful contribution. He almost never lent his name to any enterprise unless he could plan to spend a reasonable amount of time on it, and he never was a figurehead for any group.

Most of these outside activities were valuable in one way or another in furthering his fundamental objective: the progress of the Illinois chemistry department and of the university in general. Adams's capacity for hard and efficient work enabled him to direct a large research group, lead a large chemistry department, and carry a heavy burden of additional activities, most of them outside Urbana. His scientific research and his work in the scientific community were recognized by many honors.

As an active and energetic participant in his professional organization, the American Chemical Society (ACS), Adams furthered its concerns for many years.[1] He served from 1923 to 1932 as associate editor of the *Journal of the American Chemical Society*, edited for many years previously by W. A. Noyes and during this period by Adams's former Harvard colleague, Arthur B. Lamb. Adams was a member of the board of directors of the society from 1932 to 1934 and from 1941 to 1949 (chairman, 1945–49) and was president for 1935. This prestigious office carried many obligations and many opportunities for Adams to enjoy his natural talent for leadership. Contacts with the local sections throughout the country were a pleasant duty; they involved many visits and lectures personally and scientifically inspiring to chemists and students, useful to Adams in his assessment of the nation's scientific capabilities, and appealing to his effervescent enjoyment of new friends and new places. A typical trip to Montana State College at Bozeman drew a commendation from its president, A. Atkinson, to President Willard of Illinois, emphasizing the excitement and inspiration kindled in his students by Adams's visit, even if the time away disorganized his departmental work at Illinois.[2] Working for several years on the ACS committee later known as the Committee on Professional Training (CPT), he kept in touch with undergraduate training in chemistry around the country, since this group set standards and published a list of approved schools, all visited by a committee representative.

Adams devoted much time to the editorial board of *Organic Syntheses* from 1920 to 1933, when the first twelve annual volumes were issued, and after that he served on the advisory board. The term "advisory" is misleading; on *Organic Syntheses*—his favorite project outside of research—he "advised" on all matters, from investments to the amounts spent on dinners for the editorial board of *Organic Syntheses*. Even in his last years he followed this publication with alert care and never hesitated to make his views known. He had conceived the idea of the publication and had made it a worldwide force in organic chemistry, and his continuing active interest in its progress was natural.

The records of *Organic Syntheses* before 1937 are evidently lost; the editorial board apparently met as in later years for an afternoon of work and dinner at each national ACS meeting. From 1928 to 1937 (volumes 9–17), C. F. H. Allen, one of Kohler's Ph.D.'s then teaching at McGill, acted as secretary. In 1938 A. Harold Blatt of Queen's College, another Kohler Ph.D., became secretary and was associated with *Organic Syntheses* or *Organic Reactions* for many years. Until 1939 the enterprise proceeded very informally, evidently along Spartan lines financially, with Adams keeping the accounts. Thus Adams wrote Blatt in 1937:[3]

> I am enclosing herewith a check for $100.00 for your expenses in connection with "Organic Syntheses." I suggest that you keep your accounts and shortly before this amount is expended you can send me the statement and I will forward you additional funds.
>
> I was very happy indeed to know that you are willing to act as secretary of our group. I feel that "Organic Syntheses" is a complete success, but it can continue so only if we have efficient people associated with it.

Further correspondence between Blatt and Adams revealed that the former's expenses as secretary were mainly for postage. Adams and other editors invested the royalties from *Organic Syntheses*, and no one received any salary, although the secretary did receive travel expenses ($54.00) for attending the ACS meeting in Rochester in 1937.[3]

In 1939 a change in the income tax laws led to the incorporation of *Organic*

Syntheses as a nonprofit membership corporation in New York State.[3] The officers of the corporation were as follows: Adams, president; W. W. Hartman, treasurer (Eastman Kodak); A. H. Blatt, secretary; L. F. Fieser (Harvard); J. R. Johnson (Cornell). Thereafter the corporation held its official meetings separately from the board of editors. The latter normally numbered about six, and each member was responsible for assembling at least one of the annual volumes. Adams's other innovation in chemical publications, *Organic Reactions,* was first discussed apparently in 1939,[4] but volume 1 did not appear until 1942.

Adams wrote *Elementary Laboratory Experiments in Organic Chemistry* with his former student, John R. Johnson. First issued in 1928, revised editions were published in 1933, 1940, 1949, 1963, and 1978 so that it remained in print for over fifty years and sold many thousands of copies.[5] The revisions of the later editions were largely done by Johnson with assistance on the fifth edition by Johnson's younger colleague at Cornell, Charles F. Wilcox, Jr. (b. 1930). The longevity and wide use of "Adams and Johnson" rivals that of the classic Gatterman–Wieland laboratory manual, which was translated into English and reprinted repeatedly. Conant had high praise for Adams's book in 1928.

By the 1930s, the factual material and also the development of ideas on reaction mechanisms had expanded enormously in organic chemistry, but there was no satisfactory advanced text for students or research workers. W. Hückel's stimulating two-volume treatise in German appeared in 1931, but the language made it difficult reading for most graduate students. Henry Gilman, now well launched on his monumental career of research on organometallic compounds and heterocycles at Iowa State, undertook the preparation of a collaborative treatise by American organic chemists, giving authoritative summaries of important topics. Roger Adams and C. S. Marvel not only served on the editorial board but also, with R. L. Shriner, contributed a 250-page chapter on stereoisomerism that represented a thorough summary of classical stereochemistry. R. C. Fuson wrote an excellent chapter on alicyclic compounds and the strain theory that is still useful as a summary of current ideas and experimental knowledge. Thus the four senior organic chemists at Illinois all took part in Gilman's *Organic Chemistry: An Advanced Treatise.*[6]

The Gilman *Treatise* appeared in two large volumes early in 1938 and enjoyed great success because it filled a real need. A second edition, which contained some new and omitted some old chapters, appeared in 1943 and a third edition was published in 1953.[7] The makeup of the editorial board was similar to that of the early volumes of *Organic Syntheses* and, together with the contributing editors, represented the Establishment in organic chemistry of the time, dominated by Illinois and the midwestern state universities, with representation from Harvard, Columbia, Princeton, Caltech, and other laboratories.

Adams's election to the American Academy of Arts and Sciences in 1927, then a rather provincial Boston organization founded by John Adams and others in 1780, inaugurated the cavalcade of awards and elections to honorific societies that recognized his scientific accomplishments. The Nichols Medal of the New York Section of the American Chemical Society followed in 1927, and in 1936 the Gibbs Medal of the Chicago section.[8] He enjoyed admission to the American Philosophical Society in 1936 and traveled to its annual meetings occasionally.

Election to the National Academy of Sciences in 1929 at the early age of forty was a distinct honor. He was a member of its council from 1934 to 1937[9] and chairman of the section of chemistry from 1938 to 1941. The chemistry section had only about thirty-five members at that time, and the chairman's duties were not as burdensome as they later became.[10] Adams served the academy in various capacities during later years; in 1934 President Roosevelt appointed him a member of its short-lived Science Advisory Board, which proved unable to mobilize scientific advice to ameliorate the Depression.[11] Although the board accomplished little, it was useful experience for those like Vannevar Bush, K. T. Compton, and Adams, who organized the National Defense Research Committee and related groups for war research in 1940. It also made Roosevelt and his advisors aware of the scientific and administrative talent available in the National Academy of Sciences.

In 1933 James B. Conant was chosen president of Harvard from his post as chairman of the chemistry department, and because E. P. Kohler was close to retirement, a senior position in organic chemistry opened up at Harvard. Kohler, then chairman, offered the position to Adams in a remarkable letter written January 23, 1934:[12]

> It was twenty years ago, was it not, that while advising you to accept the offer from Illinois, I predicted that sooner or later we would call you back—rich in experience and full of honors. Much of this prediction has long since been verified in fullest measure. It is now my very great privilege to try to make the rest come true.
>
> Here it is. We invite you to become Sheldon Emery Professor of Organic Chemistry with a salary of $12,000, the top salary at Harvard, to take over as many of Conant's activities as will be feasible.

Kohler hoped that Adams would continue his work on natural products and develop common interests with some of the biologists who were attempting to set up a viable program in biochemistry. Kohler continued:

> There is the picture as I see it. I know that it will not be easy for you to decide. Modesty forbids that I should dilate on our merits and pride that I should enlarge on our faults. You know them both because more than anyone else you have been in intimate touch with us throughout the years that you have been away. And I think you know that if you decided to accept our invitation it would be coming home, not like a prodigal son, but like one returning with much honor from far places.

On the same date Conant wrote Adams emphasizing the possibility of collaboration with the biochemists and biologists at Harvard and expressing his personal

Illinois University Archives

Consulting at DuPont, with Dr. C. W. Tullock (1953).

James B. Conant (1934).

Harvard University Archives

hope that Adams would accept. Although Adams considered this offer to leave Illinois about as seriously as he considered any he received, and Mrs. Adams very much wanted him to accept it, he eventually decided against it. As usual, there is no evidence from him about the reasons for his decision. However, the following letter to Dean McClure on February 19, 1934, written longhand by Adams himself, shows him playing poker with the Illinois administration. It would be hard to find a better example of the iron hand in the velvet glove. This letter also names some additional activities of Adams with the National Research Council:[13]

> In considering the question as to whether I shall remain at Illinois or accept the Harvard position, there have been a few points in connection with my present position which I should like to raise. During the past years my contacts with outside scientific organizations have continued to increase. I am referring to such appointments as my directorship of the National Academy of Sciences, membership in the General Executive Committee of the National Research Council, membership on the National Research Fellowship Board, service on several national medal committees, in particularly the Willard Gibbs, T. W. Richards and Langmuir awards, and my duties on two editorial boards. For the past few years I have been a director of the American Chemical Society and during 1935 am to be its president. In this latter office I shall have to devote considerable time to the organization. I recognize, of course, that I could eliminate some of these items but they are not only an advantage to me but also I believe to the University. The duties on committees of the University and in the chemistry department have by no means decreased and I find it more and more difficult to keep up with what I am expected to do and want to do.
>
> What I need especially to help me out is the aid I could receive from two men of the rank of instructor at about $2,000 each. One, I would use as research instructor, take my classes when I am out of town and especially help in the initial. . .instruction which all the new research students who come to study with me must have. When he is not busy with these services, I should like to use him either in connection with certain general departmental affairs or as a research assistant. The other man I would use as a research assistant for full time. My intention would be to retain these men for perhaps two years and then replace them at the same salaries with other competent young Ph.D.'s.
>
> I have been handicapped in comparison with colleagues who are professors of organic chemistry in other universities, Massachusetts Institute of Technology, Harvard and Johns Hopkins I have particularly in mind. They have each had at least one full-time Ph.D. research assistant although they have had fewer executive duties and no more general teaching than I have had. Conant at Harvard had five or six such assistants which gave him a tremendous advantage over the rest of us in the type of problems he could attack.
>
> May I add that I firmly believe the department of chemistry at the University of Illinois will have a bright future provided it can be held reasonably well together and in particular if some of the senior men can be retained. There are about four or five of our members who are likely to receive offers from time to

time from other organizations and I should have confidence in the future if I could be assured the University would go to the limit to retain such men.

You realize that the present decision I must make is for me personally a major one and consequently I feel I must weigh carefully all the factors involved. It is for this reason I am troubling you with this letter for information.

The Illinois administration met his request for two postdoctoral fellows[14] and Adams declined the Harvard offer. The following extracts from a letter of March 15, 1934, to Adams from Kohler show the Harvard reaction:[15]

> Naturally we are disappointed at your decision but equally naturally we were not altogether surprised because we realized how deeply rooted you are in Illinois. Also, being realists, we have no delusions—or few—about the drawing power of "Harvard's prestige." We counted much more on the comparative freedom from administrative routine that we enjoy. I now console myself with the thought that you are not yet old enough to enjoy being coddled by a younger generation which thinks it is managing you while it serves you. . . .

Adams undoubtedly wanted to stay at Illinois to continue to head the highly successful department there. He was familiar with the Harvard scene and certainly enjoyed the less formal life in the Midwest. He may have agreed with his distinguished distant relative, Henry Adams, who described Cambridge and Harvard of the 1870s as having a social climate that would have starved a polar bear.[16] He probably felt that he was accomplishing more at Illinois than he could at Harvard. Furthermore, he was president-elect of the American Chemical Society in 1934, and moving before 1936 would have offered severe practical problems.

The English chemist R. P. Linstead filled the position at Harvard for a brief time. Soon after the outbreak of World War II he returned to England, and thereafter organic chemistry at Harvard was in the hands of L. F. Fieser, P. D. Bartlett, and R. B. Woodward; Kohler died in the spring of 1938.

The Harvard offer was scarcely settled when Karl T. Compton, president of MIT, journeyed to Urbana to invite Adams to go to MIT, at a salary of $12,000 with $10,000 additional the first year to pay for expenses of relocation. Adams declined this offer also, with thanks.[17] By this time the Illinois administration was convinced of Adams's preeminence and treated him with caution and marked respect. In 1936 Adams had three postdoctorates supported by the university; whether the third one was the result of another academic offer is not known.

Adams had a nominal connection with the National Farm Chemurgic Council.[18] Founded by William J. Hale (1876–1955), a competent organic chemist at Michigan and then director of research at Dow Chemical Company, and by Francis P. Garvan (1875–1937), head of the Chemical Foundation, this group aimed to use agricultural products as raw materials in industrial chemistry. Farmers were not generally prosperous even in the 1920s, and in the Great Depression of the 1930s they suffered from ruinous surpluses of crops as well as from dust bowl conditions in some areas. The windy rhetoric and slightly crackpot air of the chemurgic movement were completely foreign to Adams's mode of thought. However, the Chemical Foundation generously supported chemical research at Illinois and elsewhere, and Adams, who never forgot that he was head of a chemistry department well supported by tax funds from a great agricultural state, felt obliged to join a group that hoped to find new, profitable uses for farm products. Adams was asked to join the group because of his high standing in chemistry and his chaulmoogric acid work, for Garvan was interested in chemotherapy.

The Farm Chemurgic Council held several national meetings at Dearborn, Michigan, in which Henry Ford participated prominently. Garvan's death took away some of its motivating force, and the approach of World War II diminished its importance. In spite of its eccentricities, the chemurgic movement suggested some ideas that received detailed examination many years later in an entirely different social and economic setting. These included the production of ethanol by fermentation for use in gasoline and the general idea of utilizing solar energy incorporated into plants ("biomass") as a source of energy.

The most valuable and concrete result to which the chemurgic movement contributed was the establishment of four regional research laboratories by the Department of Agriculture, located in the Peoria, Philadelphia, New Orleans, and San Francisco areas. These gave important assistance, both theoretical and practical, to the utilization of farm products and to other problems, such as the manufacture of penicillin. Adams realized early that the chemurgic movement was not likely to lead to much, and his participation in it was limited; he is mentioned only once in Borth's book.[18] The Chemurgic Council existed through the 1950s and 1960s, and Adams served on the board of directors, resigning in 1964 because he had been able to attend only one meeting in the previous five years.[19]

Adams's career as consultant to industrial research laboratories started with Abbott Laboratories in 1917, as noted above. Ten years later a second important consulting connection began, and one of his graduate students, W. H. Lycan, recalled the event:[20]

> Roger spent the summer of 1927 at the Experimental Station [of Du Pont] in Wilmington working in large degree with Doc's [Carothers's] newly founded fundamental research section. He made heavy contribution to the directions in which the work developed and he was as completely at home with Julian Hill, Frank Van Natta, Don Coffman and others in the section as he was with his own students back in Urbana. Incidently, those of us back in Urbana were required to write weekly reports during his absence. The comments which unfailingly returned served not only to insure our attention to business but to instruct us in the art of writing plainly and to the point.
>
> A year and a half later, I was to go to Wilmington and though I was stationed across the river at Jackson Laboratory, I quickly resumed my friendship with Carothers. In the following two years...we spent many hours together. I

Illinois University Archives

Past Presidents of the American Chemical Society, 1937. Rear: Adams, A. B. Lamb, J. F. Norris, E. R. Weidlein, Irving Langmuir, C. H. Herty, Edward Bartow. Front: Leo Baekeland, William McPherson, M. T. Bogert, W. A. Noyes, C. L. Reese.

watched the development of the polychloroprenes and the polyamides at first hand. Thus I know that Roger contributed immeasurably not only as a consultant but via personal communication. I doubt there is written record of their close association at the time but there are many who can testify to it.

Adams and Marvel became consultants to Du Pont in 1928, a tie that lasted for Adams for nearly forty years and for Marvel for over fifty. Although Du Pont wanted Adams to visit every month, he decided that he could not spend the time required and suggested that they retain both Marvel and him, each to visit on alternate months.[20] Adams also consulted for A. E. Staley Company of Decatur, Illinois, who manufactured cornstarch and related products, and for several years he was a consultant to the Coca-Cola Company, and to the M. W. Kellogg Company. These latter associations are not nearly as well documented in his papers as his Abbott and Du Pont arrangements, which lasted much longer.[21] In all or most of these laboratories there were students of Adams or of someone else at Illinois. Indeed, it was a small research laboratory that by the 1930s did not have at least one Illinois chemist on its staff. Adams's visits were social as well as business occasions and allowed the Illinois graduates a chance to hear about Urbana and about their contemporaries. Adams's personality had a tonic and invigorating effect on all the people he talked to, and he was welcomed as heartily by the non-Illinois chemists as by the Illini.

Entirely aside from the social aspects of his consulting visits, Adams furnished solid benefits to his consultees. He had an encyclopedic knowledge of synthetic and structural organic chemistry, including recent developments that had not yet reached the scientific journals, such as the availability of new compounds from industrial exploratory research, and he was unusually well informed about economic and political trends that might affect a potential industrial process. He took his consulting seriously; he frequently wrote the chemists involved about papers he had read that might be useful to them. Without breaking confidence among his various consulting connections, he could often say, "Somebody told me a while ago about using this reagent for this purpose," or some similar information. (He did the same for his research students in Urbana.) He also was able to recommend good prospective candidates for research positions from the Illinois department or from other laboratories. In addition to his other functions, Adams was in fact a one-man employment agency whose opinion about chemists was widely sought; Marvel was equally influential in his recommendations.

The benefits to the Illinois department from Adams's consulting activities included research or fellowship grants from industry, an excellent choice of whatever jobs were open for graduating students, at bachelor's, master's, or doctorate level, and early information about new chemicals, reactions, and instrumentation. A more subtle but nevertheless important result was the feeling in industry and in many other universities that the Illinois laboratory was at the center of the chemical world. Although this was true only with some qualifications, Adams and his colleagues had an extraordinary influence on the development of chemical research in this country, and they had a comprehensive knowledge of industrial research.

Adams, as always, had a carefully thought-out view of the benefits of his consulting activities:[21]

> ...The professor will broaden his viewpoint [by industrial consulting] and appreciate better the character and significance of investigations in industry. He becomes acquainted first-hand with current industrial problems and is thus in a better position to blend into his lectures the basic chemistry most desirable for the student. He can describe to the student how an industrial laboratory operates and can be of more help to the industrial personnel representative by indicating to what type of position the qualifications of a particular student are most suited. He learns new techniques and much new chemistry which not infrequently may contribute to the solution of his academic problems. He sometimes encounters fundamental problems which the industry does not care to undertake and which are of sufficient general importance to be investigated at the university. The fact that his wider knowledge and experience can be of help to bench chemists and technical executives in industry is a lift and stimulus to the professor. A few days away from his normal routine, occasionally, is helpful to any person and decidely so to a professor. Sometimes he can combine a consulting trip with a scientific meeting, whereas he could not afford to go if he had to pay all of his expenses. Finally and of no minor importance is the additional compensation to supplement his university salary.
>
> What can the professor contribute to industry? What he cannot do should perhaps be mentioned first. He cannot be expected to contribute novel ideas for new products which will become of industrial significance. The professor's knowledge in the general field in which he is consulting is broader with respect to the old and current literature than that of most industrial men with whom he will discuss chemistry and who usually become experts in narrow fields. Consequently he can often present a fresh and different point of view to the problem under discussion and bring to bear on the problem certain fundamental data that have not been considered previously. He may through questions stimulate the chemist's thoughts into additional channels. A new approach may be suggested. Such discussions may frequently lead to the acceleration of a problem or of equal importance to its discontinuance at an earlier date than otherwise might have occurred. The industrial man can get an impartial opinion as to why a new problem under consideration should or should not be undertaken.
>
> But I believe the consultant contributes more than just technical knowledge. His presence convinces the industrial chemist that the company has a real interest in science per se. He is usually a stimulus to the men in the laboratory. A teacher–student relationship often exists between consultant and chemist and provides a means for discussion of personal problems whether they be private or professional. Chemists will often mention to the consultant minor or more serious complaints about the laboratory or its management when he would not do so to his official superior. Consultants can keep the higher level

executive informed of any lowering of morale so that it may promptly be corrected. An opinion by consultants of the qualifications of its employees may at times be sought. Executives may acquire information of scientific trends in this country and abroad.

As the number of Adams and Illinois Ph.D.'s grew, the influence of the Illinois department and its graduates became very great, particularly as many of these graduates reached responsible administrative positions in research and teaching. The number of Illinois graduates in many organizations was sometimes discouraging to those who lacked this background, because they felt, correctly or not, that their chances of advancement were slighter. Adams himself had a remarkably objective view of other people (and himself); although between two people of equal ability, one an Illinois graduate and the other not, he would pick the Illinois man, he would not take an Illinois degree as a substitute for ability and performance.

Adams always regarded himself as a member of the academic profession, yet with his natural inclination toward the business world, he spoke the language of businessmen and was liked and trusted by them. His firsthand knowledge of finance and of management in general grew with the years. Adams frequently wrote and spoke about what industry was entitled to expect from its research chemists, and the majority of Illinois Ph.D.'s who did go into industrial research were well informed as graduate students in this regard. Conversely, Adams was able to make clear to research management in industry how research chemists should be treated to maintain good morale and productivity.

Naturally his collaborators speculated about his income from consulting. The documentation, although not complete, shows that his consulting fees were far smaller than popularly believed, and that the department and university probably benefited more than he did personally.[21] The outside income helped to make up for the very large amount of time and effort he devoted to public service activities.

Early in his career Adams joined the American Association for the Advancement of Science (AAAS) and rose to prominence in the organization, becoming president in 1950. A very important accomplishment for American science was his success in persuading the AAAS to sponsor the useful and stimulating Gordon Conferences. Started by Neil E. Gordon (1884–1949), professor of chemistry at Johns Hopkins, to examine chemical topics and first held on Gibson Island in Chesapeake Bay in 1931, they soon outgrew this site. Adams tried to have the American Chemical Society sponsor them, but the board of directors refused, because some of them disliked Gordon, in Adams's view. He prevailed upon the AAAS to take over the meetings in 1938, a logical arrangement, as the conferences soon expanded into many areas beyond chemistry.[22] They were moved to New London, New Hampshire, and they are now held every summer at numerous locations in New Hampshire and on the West Coast.

LITERATURE CITED

1. H. S. Skolnik and K. M. Reese, eds., *A Century of Chemistry*, American Chemical Society, Washington, D.C., 1976, passim. For the Committee on Professional Training, pp. 65–71.
2. A. Atkinson to A. C. Willard, November, 29, 1935, Illinois Archives 2/5/15, 364.
3. Voluminous correspondence about *Organic Syntheses* in RAA, 25 and 26, and in the secretary's file of *Organic Syntheses* demonstrates the affectionate and detailed care that RA devoted to the publication; RA to Blatt, May 13, 1937; *Organic Syntheses* records; R. L. Shriner and R. H. Shriner, *Organic Syntheses, Cumulative Indices*, Wiley, New York, 1976, pp. 423–32; RA to Blatt, June 5, 1940.
4. A. H. Blatt to RA, January 8, 1940, giving early plans for *Organic Reactions; Organic Syntheses* records.
5. RA and J. R. Johnson, *Elementary Laboratory Experiments in Organic Chemistry*, Macmillan, New York, 1928.
6. The RA correspondence furnished to us by Henry Gilman deals mainly with Adams's work on the third edition of 1953.
7. *Organic Chemistry, An Advanced Treatise*, editorial board, Henry Gilman, editor-in-chief, RA, Homer Adkins, H. T. Clarke, C. S. Marvel, and F. C. Whitmore; 2 vols., Wiley, New York, 1938; second edition, same editors, 2 vols., 1943; third edition, 4 vols., 1953. RA was still on the editorial board and took an active part.
8. RA was introduced for the Gibbs Medal by A. C. Willard, president of the University of Illinois, in a speech that has some useful details; RAA, 4, 1936–1939.
9. R. C. Cochrane, *The National Academy of Sciences: The First Hundred Years 1863–1963*, National Academy of Sciences, Washington, D.C., 1978, pp. 614, 642.
10. From Reports of the National Academy, 1937–1938 to 1940–1941, courtesy of the archivist, Jean R. St. Clair.
11. For this advisory board, see Cochrane, op cit., pp. 346–81.
12. RAA, 64, Correspondence 1913–1964; E. P. Kohler to RA, January 23, 1934; J. B. Conant to RA, *ibid.*
13. RA to Dean M. T. McClure, Dept. and Subject File, Dean's Office, Illinois 15/1/1, 27, Chemistry, 1933–1934.
14. Conant was the first American chemist to utilize postdoctoral fellows systematically; he obtained the funds for them from the Harvard administration as the result of an attractive offer of 1927; Conant, pp. 73 ff.
15. RAA, 7, Alphabetical Correspondence, 1912–1971; E. P. Kohler to RA, March 14, 1934.
16. Henry Adams, *The Education of Henry Adams*, Ernest Samuels, ed., Houghton-Mifflin, Boston, 1973, p. 307.
17. RAA, 64, Correspondence, 1913–1964; K. T. Compton to RA, longhand, apparently written on the train back to Boston, March 14, 1934, typed letter to RA, April 6, 1934.
18. Christy Borth, *Pioneers of Plenty, The Story of Chemurgy*, Bobbs–Merrill, Indianapolis, 1939, p. 166. This book is a highly colored journalistic account of Garvan, Hale, and Herty's work on producing wood pulp from southern pine, the use of alcohol in gasoline, the introduction of soy beans to American agriculture, the work of George Washington Carver, and similar topics, including the discovery of sulfa drugs. Hale wrote numerous books, such as *Farmer Victorious: Money, Mart and Mother Earth*, Coward-McCann, New York, 1949, which is cranky and virtually unreadable. A more

factual account of the chemurgic movement by Hale is in Williams Haynes, *American Chemical Industry*, Van Nostrand, New York, 1954, Vol. 5, pp. 486–90; also ibid., pp. 141–45; C. W. Pursell, Jr., *Isis*, *60*, 307 (1969).

19. RAA, 21, Chemurgic Council and National Farm Chemurgic Council, 1951–1957.
20. Adams at Du Pont: W. H. Lycan to DST, June 12, 1979; consulting for Du Pont: taped interview by DST with C. S. Marvel, March 6, 1977.
21. RAA, 16, contains extensive correspondence with Abbott about consulting; also RAA 27, Abbott; RAA 33, Du Pont, RA to R. M. Joyce, January 23, 1964; RAA 21, Coca-Cola, 1950–1960; RAA 32, Coca-Cola, 1961–1968. Adams's view of consulting is from his Perkin Medal address, *Ind. Eng. Chem.*, *46*, 510 (1954).
22. RAA, 19, Battelle, 1967–1968; RA to S. L. Fawcett, October 21, 1967. Adams's work in this connection is not generally known, but it constitutes a major service to American science.

Europe, 1936

After his strenuous year in 1935, in which the presidency of the American Chemical Society was added to his normal responsibilities, Adams took the summer of 1936 for an extended trip to Europe. He was one of twenty American delegates to the Twelfth International Conference on Chemistry, held August 16–22, 1936, in Lucerne and Zurich. It was his first trip to Europe since his year in Berlin in 1912–13. His wife and their daughter Lucile, who was nine at the time, did not accompany him, and his experiences are described in lively detail in a series of letters to his wife.[1]

Adams sailed on the North German Lloyd *Columbus* in June and returned September 2 on the *Bremen*. During the trip Adams became well acquainted with many of the leading European chemists, and he renewed his friendship with Leopold Ruzicka of the Federal Technical Institute (or ETH) in Zurich, who had visited Adams in Urbana in 1935. Wallace H. Carothers, Adams's brilliant student at Du Pont, whose psychological problems were becoming ominous, joined Adams in Munich and went for a hiking trip in the Austrian Tyrol with Adams and James F. Norris of MIT. Norris was an old hand at both international meetings and hiking in Austria. Carothers evidently was benefited greatly, although unfortunately only temporarily, by the outing.

Adams landed at Cobh, Ireland; he wrote, "It was rather a shame for the group at the Captain's table to break up, as everybody was so congenial." He got a ride to Cork with the "countess," apparently an acquaintance from the Captain's table, and thus avoided a ride in a "little old dirty train" which ran from Cobh to Cork. Adams toured in Ireland, visiting Dublin, and then crossed to England, where he visited Oxford.

Adams was struck by the English gardens; the home of a friend was

> very attractive with a beautiful garden front and back—surrounded by five foot walls which is the ordinary thing here. The back garden which is a rectangular spot must be 125 ft. deep and about 50–60 ft. wide—perfect grass in the center and a flower border all around—you have never seen such roses, delphinium 8 ft. high with blossoms 3 ft. long, lupins, flox [sic] and many others.

He commented on how immaculate the house was, the lady of the house is "like yourself in wanting everything just so." In addition to Oxford colleges, Adams visited the chemical laboratories, talked with the famous organic chemist Sir Robert Robinson for an hour, and chatted with the distinguished inorganic chemist N. V. Sidgwick, "a grand fellow," who was in the hospital. Oxford was the first place he had seen that he thought would be a nice place to live—"beautiful and so different." He did not visit Cambridge.

In London Adams met "Keis" (E. K. Bolton), who told him among other things that Carothers was in an institution in Philadelphia and was to be sent elsewhere "for treatment which they hope will cure him completely (6 months or so)." Various visits in London preceded a tour of England, and a long letter to his wife from Bath described some of it. "I didn't write you [from London] I guess but I did Lucile, thanking her for the candy—it was delicious and very acceptable although not necessary to supplant [sic] the food. I'd heard about Nieuwland on the boat—very sad—he was a very good friend of mine." Father J. A. Nieuwland, a distinguished chemist of Notre Dame, had died suddenly of a heart attack at a meeting in Washington.

Adams as a tourist was more interested in people than cathedrals, although he looked dutifully at the "sights." There was a trace of the skeptical irreverence of Mark Twain in *The Innocents Abroad* in his makeup. In Salisbury, for example, he did not get up at three in the morning to see the sun rise over Stonehenge.

> I passed that up for the second trip at 6:30. Stonehenge was impressive as it dates back to 1700 B.C. and is so located and in such condition it can't be fake. How it was ever built in those days is a mystery—such high stones 25 ft. × 8 × 4 and all shaped to fit each other.

In Bath a sight-seeing trip

> ... ended at about 12 at the Guild Hall where Lord—(God knows who he is) an attractive young fellow of about 28 welcomed us.... It's cathedral after cathedral, and two more tomorrow. They are beautiful and impressive but one gets fed up. This west part of England is famous for them almost alone. The country side is beautiful but not scenic in the usual sense of the word. That's why most Americans go to the Lake region and pass this up. It's the Shakespeare district tomorrow.

During his crossing to the continent he encountered several groups of American tourists, but "fortunately all the tourists took a different one [train] from the one I was going on. I can see how Americans get the bad reputation they have, for they were pretty noisy." Luckily, on his train to Heidelberg a fellow traveler turned out to be the minister of education in Holland. "We talked about the Holland chemists whom we both knew and then on general education matters and politics. I spent a couple of hours with him. It was a real break for me." As always, Adams delighted in learning from people he met.

In Heidelberg he met the eminent chemist Karl Ziegler and his family, who had earlier visited the Adamses in Urbana. He also saw Karl Freudenberg of Heidelberg and Julius von Braun of Frankfurt, both well-known chemists; he had seen the latter during his European year. Regrettably he missed seeing Willstätter in Munich.

Throughout this series of letters there is an undercurrent of uncertainty and concern about Carothers. First he was institutionalized; then he was "much better"; then he was thinking of joining Adams ("haven't heard whether Carothers is coming"); later, "if Carothers comes, we'll plan something out." Finally, in Munich "Carothers showed up. He looked rather dejected—had been drinking heavily but I think the walk with Norris and me will help him." Adams and Norris looked forward to a hiking trip of about two weeks, before the conference started in Lucerne.

The trio went to Innsbruck, Austria (the Nazis did not take over Austria until 1938), which they planned to use as "head quarters and go out for probably two five day trips before going to Switzerland. Carothers appeared more optimistic last evening and even more so this morning so I hope by a couple of weeks he will be feeling more optimistic."

An interesting and vivid letter from Innsbruck on August 15 describes their hiking trip. They crossed into Austria at Kufstein, south of Munich and only a few miles from Berchtesgaden, Hitler's mountain retreat. Border-crossing formalities from Germany at that time included troublesome and complicated currency restrictions. At Innsbruck "we put up at the Europa Hotel, about the 2nd or 3rd best—running water in room (7 Aust.[rian] sch.[illings] about $1.40). After dinner the three of us talked over plans—we decided to go to Stubaital first then formulated the general policy not to decide what to do the following day or morning until we had finished breakfast at the hour desired by each and had smoked two cigarettes apiece." Stubaital is a valley southwest of Innsbruck.

The trip followed the informal course outlined. They hiked or sometimes traveled by bus, carrying light packs, and stayed at inns in the small towns.

> It's difficult to describe these Inns—many people in them were Austrians spending a 2-weeks vacation—some French—an occasional Englishman and no Americans—not quite elegant enough I guess and only the landlady knew a few words of English. . . .We complimented the cook on the dinner served and so about nine o'clock, the manageress, the sister and the two cooks entertained the three of us in the Bürgerstube, the room in each Inn especially arranged for natives. The cook played the guitar and sang and we conversed till time to go to bed.

Carothers and Adams made their hardest climb without Norris, who was much older than the other two and wanted to take it easy for a day.

> He said he would walk down the valley and reserve rooms and we could climb to one of the Alpine Inns—they call them Hütte [huts] right at the base of the glaciers. . . . The trail is about like the ones in Glacier park though filled with rocks. . . . It is suitable for a horse to go up and all supplies to the Hütte are by pack-horse. The climb was about 3 hrs. going as fast as we could and we went up about 3500 ft. Quite ready for a rest at the Hütte where we found about 15 people who had been climbing glaciers. During lunch it snowed and blew but fortunately cleared off beautifully so we had an hour to look around.

During the hike down it rained, and in the valley they got to their destination by bus in time for dinner.

> We went to bed early and were decidedly stiff the next day—so decided after breakfast to take the train—a dinky little car built about 50 yrs. ago that winds in and out of the hills and finally gets to Innsbruck—a really picturesque ride as one could look down in the valley and get beautiful views.

On the way down they planned another foray to a region "where Jimmy hadn't been," the Sellraintal.

At Oetztal "we picked the cheapest hotel in Baedeker—an Alpen Rose and really had a wonderful time. In these places, the owner, wife and all the staff put themselves out to accommodate—no Americans—a few English French and Austrians—particularly Viennese. You find the crowd that have to travel cheaply and they are much more congenial." It is a pleasant picture of Adams and his two distinguished companions, enjoying and contributing to the gemütlichkeit of a tiny Austrian inn.

After several more days they returned again to Innsbruck. Adams calculated that the daily costs while they were hiking averaged $4 apiece; without

> ...smokes or drinks, the cost would have been below $3 a day. It was really wonderful—lots of fun, the skin all peeled off my forehead, lots of sleep, fresh air and exercise. I feel real fit—Carothers looks so if it will only last....Innsbruck is filled with tourists—mostly English and Amer. but I don't know where they go unless they merely stay in the big cities—we saw no Americans during our whole trip....Am not looking forward to Lucerne—Jimmy pictures a lot of boring banquets but I think it will probably be better than I'm anticipating. Carothers has decided to stay here [Innsbruck] and climb up some more of the valleys—I think this best too for him.

Carothers returned to the United States before Adams.

The hiking trip was an idyllic interlude with tragic overtones. For Adams and Norris, both busy and well adjusted in their professional and personal lives, it was a welcome change from their usual laborious careers, while for Carothers it was a brief respite from his troubled life. When they parted, Adams and Norris resumed their trip to the international meetings, but Carothers returned to the lonely path where no one could help him.

Adams actually found the meetings and banquets interesting, as might have been predicted, with new people to meet and converse with. He described the sessions in detail and the rest of his travels in further letters to Mrs. Adams.

The high point of Adams's summer, next to the hiking trip, was his visit to Switzerland after the meeting. He renewed a cordial friendship with Leopold Ruzicka in Zurich. Ruzicka, one of the world's great organic chemists, was famed for his work on large ring compounds, terpenes, sex hormones, and other natural products. As a European professor, he commanded the services of all the graduate students he wanted as well as a corps of highly experienced postdoctorates to prosecute his complicated research problems at ETH.

He invited Adams to stay at his handsome, modern home overlooking Zurich. Adams accompanied him on a consulting trip to Geneva, and from the air he admired the splendid panorama of the Bernese Alps. He had time to explore the city before joining Ruzicka and a company representative at his lakeside villa. Back in Zurich, Adams lunched with Paul Karrer, the eminent professor of chemistry at the University of Zurich and noted for his research on vitamins, carotenoids, and natural products. In that city Adams also encountered his student Byron Riegel (Ph.D. 1934), who had spent a year in Germany on a traveling fellowship.

Persuaded by the Ruzickas to stay with them a few more days, Adams accompanied Ruzicka to Basel, where he met his old friend Arthur Stoll, Willstätter's assistant at Dahlem when Adams worked there. Stoll, an outstanding chemist, was now prosperous as vice-president of the pharmaceutical firm Sandoz. It is pleasant to imagine Adams's enjoyment of the meetings and conversations—surely turned to natural products—with the Swiss chemists. All were at the height of their careers, the one from America and the others from Europe,[2] representing two different systems in the impressive progress of organic chemistry.

The Ruzickas were coming to the United States for the tercentenary celebration of the founding of Harvard College, where he would be one of the eminent scholars invited to present a paper and receive an honorary degree,[3] and Adams helped him with his English for his paper. Before leaving Zurich, Ruzicka arose at four in the morning to write up research work, and they all departed for Cherbourg a few days later.

Adams found the *Bremen* luxurious, the food and service excellent, but he was "disappointed that nobody gets acquainted with anybody else. The tables in the dining rooms with few exceptions are for four people. I am sitting with the Ruzickas and I don't suppose I've talked with more than three people." Congenial as he and Ruzicka were, Adams was anxious to meet other interesting people and no doubt, before the voyage was very old, managed to do so. The intimacy of his relations with Ruzicka is shown by some amusing correspondence in later years.

The end of this eventful trip drew from Adams an unusual expresion of regret; he wrote his wife that he dreaded to think of all the incidentals to finish before the university opened.

Adams's research program continued at its very high level both quantitatively and qualitatively in 1936–37. However, the spring of 1937 brought him great personal sorrow: Wallace Carothers committed suicide on April 29.

Literature Cited

1. The letters, some of them very long and written very frequently, are in RAA, 4, British Tour and European Trip Correspondence. Menus for the ocean trip both ways are in

RAA, 58. The letters are transcribed as written, except that "and" has been written out instead of using Adams's symbol. The Austrian place names, which are not always clear in Adams's script, have been checked in an atlas.

2. Within three years Karrer and Ruzicka were to share awards of the Nobel Prize in chemistry: Karrer in 1937 with W. N. Haworth, and Ruzicka in 1939 with Adolph Butenandt (though the latter declined the prize because of a Nazi decree).
3. According to *The Tercentenary of Harvard College*, Harvard University Press, Cambridge, 1937, p. 217, sixty-two scholars from all fields and countries were present and received honorary degrees; Ruzicka was one of three chemists, the others being The Svedberg and Hans Fischer. Peter Debye was also in the group, but he was apparently considered a physicist, judging from his citation.

Researches, 1927–42

The period 1927–42 covers the time from Adams's assumption of the headship of the chemistry department to the beginning of his full-time work for the National Defense Research Committee and other units engaged in scientific research to support the war effort in World War II. In contrast to his activities during his more limited time of service in World War I, during World War II he was almost completely cut off from contact with his research group, and he did not take new Ph.D. students for several years. Furthermore, after the war's end in 1945 several overseas trips on official missions consumed months of time, and although his energies could be recouped, the time could not.

Beginning about 1940 several factors, whose approach was obvious even before then, fundamentally changed organic chemistry. These changes, principally the rise of physical organic chemistry and the development of new, powerful instrumental methods, occurred just as Adams was being cut off from close contact with his research group, and under the best of conditions he would have had to study seriously to remain abreast of developments. His Washington duties prevented this, and although he carried on excellent research after 1945, he was no longer on top of the field as he was before 1942.

The 1927–42 period thus has a double interest; it shows Adams as an acknowledged master of classical organic chemistry, and it shows his work at his scientific zenith. In particular, his work on structural elucidation of natural products, including gossypol, marijuana compounds, and the senecio alkaloids, was equal to the best being done in any country. This excellence was due in part to Adams's breadth of knowledge and experience, in part to the trained postdoctoral fellows (two or three starting in 1935), and in part to the general improvement in the quality of American organic chemistry as a whole compared to that in other countries. Organic chemistry in 1940 in this country was much better than in 1920, and the Illinois laboratory contributed an important share to this improvement. In this respect, as in many others, Adams was a leader who attracted high-quality followers to the benefit of all.[1] Other leaders in natural-product work were W. A. Jacobs of the Rockefeller Institute, W. E. Bachmann of Michigan, C. S. Hudson of the U.S. Public Health Service, Karl Folkers of Merck, and biochemists associated with strong groups in medical schools, agricultural colleges, and state experimental stations. None of these, however, had as large a group of students as Illinois, and no single laboratory carried as much influence.

Some of Adams's earlier studies continued after 1926, particularly his work on the properties of the Adams platinum catalyst and the synthetic work directed to long-chain acids of the chaulmoogric acid type. A variety of long-chain acids was synthesized by many students and tested by W. M. Stanley against *M. leprae;* the activity usually peaked at 15–18 carbon atoms, and Stanley found a correlation between surface tension lowering and bactericidal action.[2] Stanley's experience in bacteriology that he developed working on this problem helped him to obtain a position at the Rockefeller Institute, where he did the research on viruses that secured him a share of the Nobel Prize in chemistry in 1946.

A problem new to Adams, which almost became his trademark and on which he published about sixty papers by 1942, was the occurrence of hindered rotation in biphenyls and related compounds. It was shown by the English chemists Christie and Kenner in 1922[3] that 6,6′-dinitrodiphenic acid could be resolved into two stable, optically active forms. This stereoisomerism arises from the fact that the large substituents adjacent to the bond joining the two benzene rings force the rings into a non-coplanar configuration and prevent their free rotation about that bond; hence the whole molecule is asymmetric.

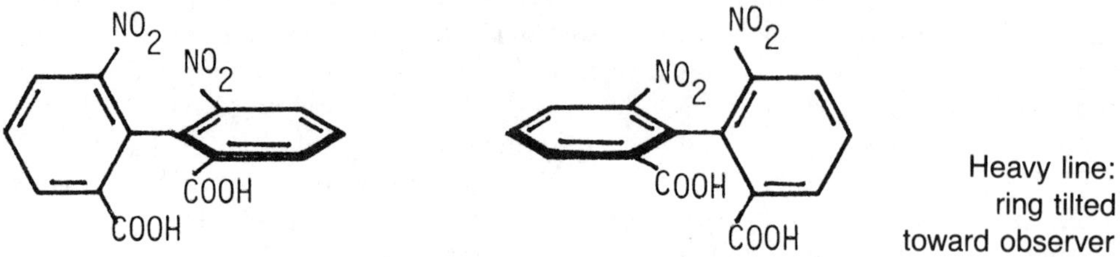

NO_2 NO_2 COOH COOH — NO_2 NO_2 COOH COOH

Heavy line: ring tilted toward observer

6,6'-dinitrodiphenic acid

Adams showed that the resolvability of biphenyls and their rate of racemization, once resolved, depended on the size and number of the groups in the ortho positions and not on their electronic character. It was possible to show a general correlation between X-ray data on the size of atoms or groups and the effectiveness of groups in preventing coplanarity and so preserving optical activity,[4] the interference radii of groups were tabulated by Adams and Yuan in a review.[5] Adams's first paper in this field was in 1928,[6] and he extended the work to diphenylbenzenes, substituted so that there were two points of hindered rotation and hence two asymmetric centers. He also studied nitrogen-containing analogs of biphenyls, such as phenylpyridyls, bipyridyls, and *N*-phenylpyrroles, finding similar evidence of hindered rotation.[7] Stanley found that two ortho sulfonic acid groups in biphenyl did not allow resolution and hence did not prevent coplanarity.[8] However, two iodine or two bromine atoms did hinder rotation.[9] It was not possible to resolve biphenyls with a single iodine or an alkylamido group.[10] Adams showed that restricted rotation did not require two ring systems by resolving the compound A below, which contained a highly substituted double bond, preventing free rotation.[11] A particularly interesting example of a biphenyl was that of Kornblum (B), where a third large ring hindered rotation.[12] Adams's colleagues in physical chemistry confirmed the non-coplanar character of optically active biphenyls by studies on the X-ray crystallography and ultraviolet absorption spectra of some of his compounds.[13]

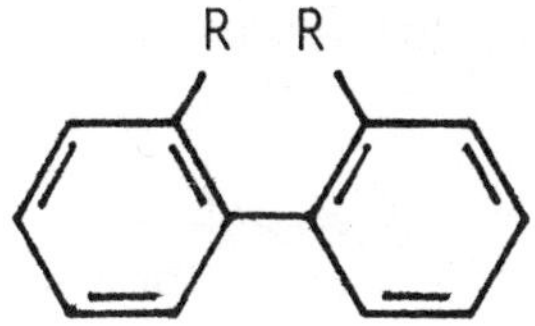

R = SO_3H, not resolvable
R = I or Br, resolvable

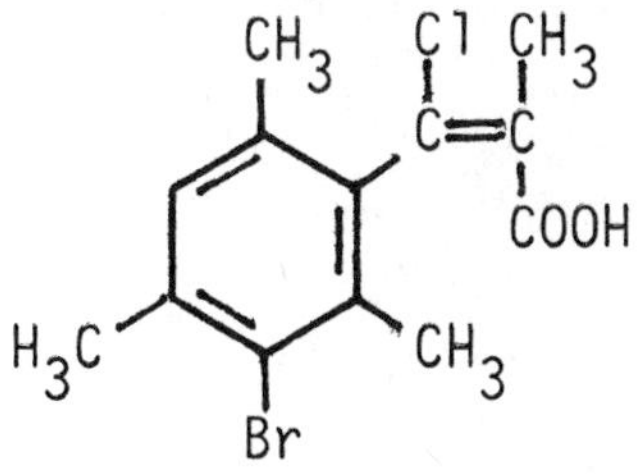

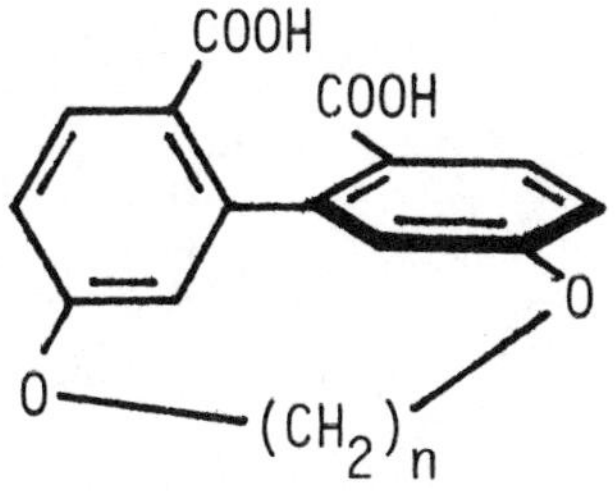

n = 8, 10

A B

As mentioned earlier, many of the chemical fraternity regarded Adams's biphenyl work, after the first dozen papers, as "busy work" and "more of the same." Adams looked upon it as a way of teaching fundamental organic chemistry, and examination of the original papers shows that this is true. The synthetic chemistry involved was frequently intricate and original, and the resolution of a racemic compound was always laborious and sometimes difficult.

A legitimate criticism of Adams's hindered rotation work is that the rates of racemization were measured only semiquantitatively; rate constants and activation energies were obtained in only a very few cases.[14] More of these would have been most useful in correlating the results on the relative sizes of groups. It is unfortunate that the Illinois department did not have a physical organic chemist to collaborate with Adams here because the optically active biphenyls were difficult to make, and many physical organic chemists would not have had the patience or synthetic skill to prepare them. Later work by Kistiakowsky at Harvard, using a compound supplied by Adams, and by Westheimer at Chicago gave activation energies of 20–45 kilocalories.[15] Westheimer also confirmed a "buttressing effect" of groups in the 3-position in raising the energy for rotation, presumably by preventing distortion of the bond angle of the groups in the 2-position. The "buttressing effect" had been observed much earlier by Adams in several cases.[16]

To the impartial observer, therefore, the hindered rotation papers of Adams contain much excellent and useful experimental work, but with better rate measurements, valuable data could have been obtained about the energies involved in groups or atoms moving past each other and in the distortion of bond angles. The problem was not formulated to furnish all the scientifically important information that was potentially there.

Adams's natural-product work during this period was superb; he tackled several very difficult problems and either solved them completely at Illinois, as in the gossypol case, or made fundamental contributions in competitive fields, as in the marijuana and senecio areas. As a warm-up, working in the 1930s with R. C. Morris and W. E. Hanford, both highly successful industrial chemists later, he established the structure of the alkaloid vasicine in 1936 and corrected earlier erroneous structures.[17]

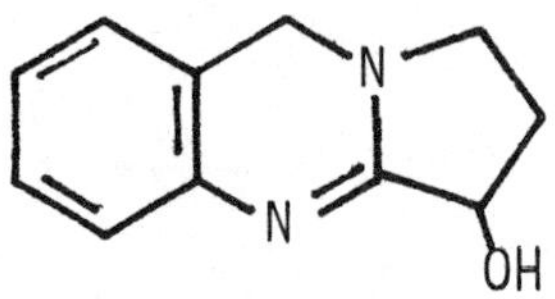

vasicine

The compound gossypol, a yellow toxic constituent of cottonseed meal, was first isolated in 1899, and several groups in government laboratories in this country and in universities overseas attempted to define its structure.[18] None of this work had made much progress when Adams started research on it in 1936 with an exceptionally able group of graduate students and postdoctorates.

Although gossypol could be obtained in pound lots by extracting 100-pound batches of cottonseed meal in equipment loaned by the Chemical Engineering Division, the compound was sensitive to base and to air oxidation and contained so many functional groups that it was soon clear why other laboratories had given up on the problem. Adams and his group persevered, working with great tenacity and ingenuity in the face of constant discouragement, and gradually they learned how to degrade the complex molecule into identifiable components. In this work ultraviolet spectroscopy was used for the first time in Adams's research, introduced by his student E. C. Kirkpatrick, who had done his undergraduate work at Princeton.

By the spring of 1938 it was possible to write a satisfactory structure for gossypol, and Adams prepared a series of 11 papers, writing the last one mainly by himself, a brilliant summary of the whole complicated problem. Adams prepared this paper in a very short time, and it was one of his masterpieces.[19] The whole series of papers was submitted to the *Journal of the American Chemical Society* at once, nearly overwhelming the editor, Arthur B. Lamb. Adams eventually succeeded in having the series published in the issue of September 1938, where it covered forty-eight pages, double column. It was said that the galley proof arrived while Roger and Mrs. Adams were at their summer home in Greensboro, Vermont, and that they read the proof together for many very tedious hours.

The binaphthyl structure shown below was proposed for gossypol, and it was supported by the synthesis of the degradation products, apogossypolic acid and desapogossypolone tetramethyl ether, by Adams and B. R. Baker.[20] A more detailed analysis of the evidence for the gossypol structure would require repetition of most of Adams's summarizing paper.[19] The final evidence for the correctness of the gossypol structure was furnished by a total synthesis of gossypol in 1958 by a former postdoctorate of Adams's, J. D. Edwards at Baylor Medical School,[21] a tour de force by a solitary worker. Many years later Chinese workers reported gossypol to have anticancer activity.

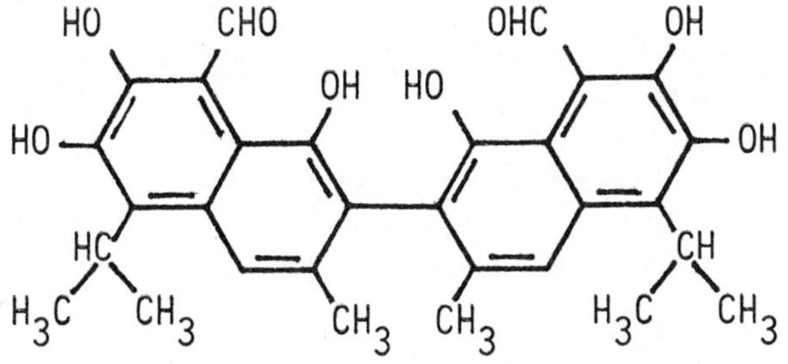

gossypol

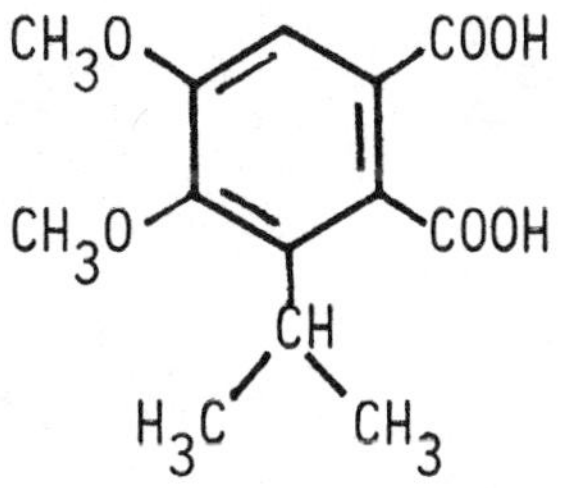

apogossypolic acid

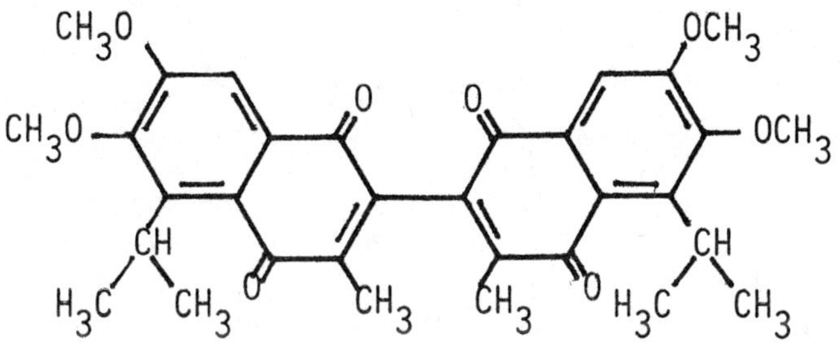

desapogossypolone tetramethyl ether

Although much cleanup work remained on gossypol after 1938,[20,22] Adams immediately started in the spring of 1939 on a chemical study of the constituents of marijuana, at the request of the Bureau of Narcotics of the U.S. Treasury Department. As is well known, marijuana comes from the so-called hemp plant, *Cannabis sativa,* which grows in this country, and also from another *Cannabis* variety, *C. indica,* which grows in the Middle East and India. Work on the chemistry of the constituents contained in *Cannabis* extracts had been done earlier in England, but although some information was gained, the problem was still not well defined and only a beginning had been made.

Adams carried out his work on the "red oil" furnished by the Treasury Department and obtained by extraction of *C. sativa* from Minnesota. He was able to isolate and prove the structures of several compounds from the red oil, of the types shown below.[23] He showed that while the cannabidiol compounds had no biological activity, treatment with acidic reagents gave oils of the tetrahydrocannabinol series, which did show some of the marijuana activity.[24] Similar work was being published simultaneously by A. R. Todd at Manchester, England, who had corresponded with Adams, and this provoked Adams to one of the very few caustic statements he ever made regarding his priority rights on a problem.[25] Later, however, he and Todd became close friends, and Adams wrote him a note of congratulations when Todd was raised to the peerage.[26]

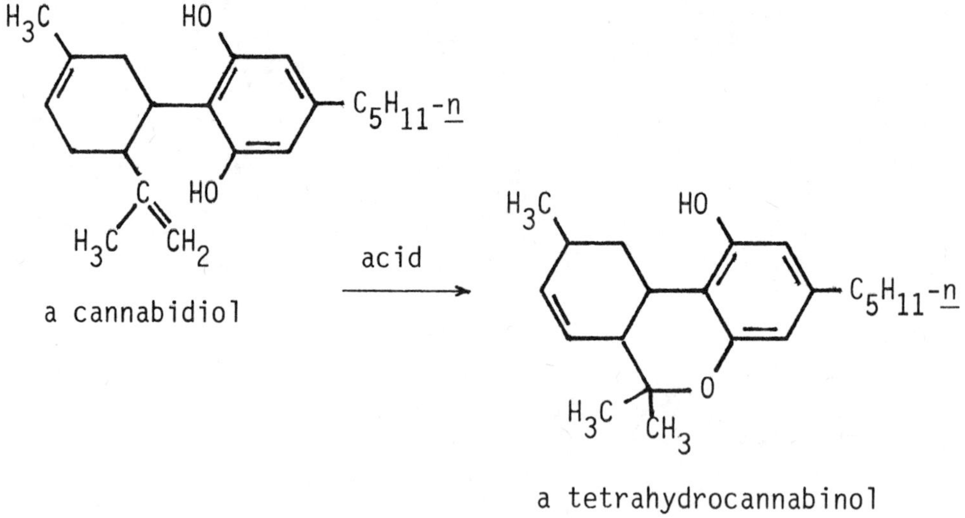

a cannabidiol

acid →

a tetrahydrocannabinol

World War II interrupted Adams's work on marijuana, and the natural active constituents of the red oil were not obtained pure until after 1945 by other research groups, when vastly improved methods of separation were available and when double bond positions could be inferred from *nmr* spectra. Adams's group did prepare a series of synthetic compounds similar to cannabidiol and tetrahydrocannabinol, and a few of his friends courageously ingested capsules of this synthetic tetrahydrocannabinol material in 1943. Several accounts of the subjective effects of the material are extant, and because they were reported by experienced scientists, one is worth quoting:[27]

> At 4:45 P.M. took one capsule (15 mg.) Pyrahexyl, went home on train, rode bicycle from station—one mile. Sat down to dinner at 6:15, remarking to my wife that I was trying an experiment with a capsule from Roger Adams, which was supposed to make one very hungry—but I did not feel more hungry than usual. Ate a good dinner, and at 7:00 rose from the table, and stood talking to my wife. Very suddenly my legs felt numb; it was 7:03. My mind felt fuzzy, disoriented. A feeling of anxiety began to develop. I paced the floor. My wife talked to me as she washed the dishes; occasionally she asked me a question, but usually by the time she had said the last word, I had forgotten the question before I could answer. Time stood still: after what seemed like the passing of hours, the clock showed that only a minute or two had gone by. My mouth was intensely dry; my tongue felt several times its usual size, but examination in the mirror showed no change. I found myself eating several pieces of candy, but had no recollection of having picked them up.
>
> By 7:15 to 7:30 I began to feel that my sanity might be impaired, and questioned whether I had really taken a capsule at all. Doubt entered my head that I would recover; my anxiety increased. To determine whether my mind was working more rapidly than usual, I turned my attention to a chemical problem which had troubled me during the day. I found that I could examine all its phases and possibilities much more speedily, but I could not solve it any better. Finally, at about 8:00 P.M., when the symptoms were at their worst, I asked my wife to phone Mr. Carter to determine whether I had actually taken a capsule in his presence at 4:45. He stated that I had told him in my office that I had done so. My wife also phoned Doctor Biehn, who, with Doctor Hazel and Mr. Carter, soon came. I talked to them freely, but not very brightly. Shortly after 8:00, Doctor Biehn gave me ¾ grain Nembutal, and I went up to bed; my gait was now rather uncertain. I went to sleep almost at once, dreamed some, but spent a rather comfortable night.
>
> I arose as usual at 6:30, feeling all right but still somewhat dull, especially in remembering things. Breakfast and appetite as usual, then I rode my bicycle to station and went to work. My memory was still poor, and I was inclined to be garrulous. At 11:00 A.M. Doctor Biehn gave me two large candy bars as an antidote, which I ate within ten minutes, and did not feel surfeited with sugar. At noon I ate my usual lunch, which tasted all right.
>
> During the afternoon the dullness and lack of memory and garrulousness gradually disappeared and by 5:00 P.M. I felt entirely normal again.
>
> I dictated several letters after lunch, and my secretary stated later that my dictation was repetitious and not very clear.

Although it was not published until 1948, Adams's group apparently found at about this time that replacement of the *n*-amyl group found in the natural marijuana products by the more highly branched chain, $-CH(CH_3)CH(CH_3)-n-C_5H_{11}$, increased its activity on the central nervous system many times. This observation has been the basis of further promising compounds made in other laboratories.

As impressive as the purely chemical work on the marijuana problem is the formidable amount of management it required. Adams's correspondence relating to marijuana has been preserved, apparently in full; he had to arrange with the Treasury Department people to get the red oil prepared and delivered, often a slow process. He also arranged to have samples sent to a European refugee pharmacologist, S. Loewe, who had received some research space at Cornell Medical College in New York and who reported the biological results to Adams. Much of this correspondence, which runs to hundreds of pages, dates from the time when Adams was stationed in Washington and could make only hurried visits to Urbana.[28] The papers show Adams's great energy in pushing his research problem and the ability with which he managed a project involving at least three research groups. Although the letters referring to the gossypol work apparently have not been saved, it is unlikely that they would show as complicated a management problem, because the gossypol was isolated from cottonseed meal in the Illinois laboratory and pharmacological testing was not part of the program.

In 1939, when the gossypol problem was nearly completely solved and the marijuana work was underway, Adams embarked on other natural-product series, the structures of the alkaloids present in *Crotalaria* and later those present in the *Senecio* species, which occupied him in part for the remainder of his scientific career. The problems involved two general parts, the structures of the acids, the "necic acids," such as monocrotic and monocrotalic acids obtained by hydrolysis, and the structure of the heterocyclic bases such as retronecine, a derivative of pyrrolizidine.[29] Alkaloids structurally similar occur in *Crotalaria*, *Senecio*, and other genera.[30]

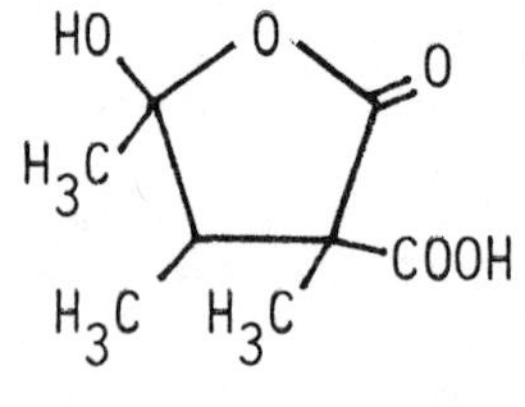

monocrotalic acid

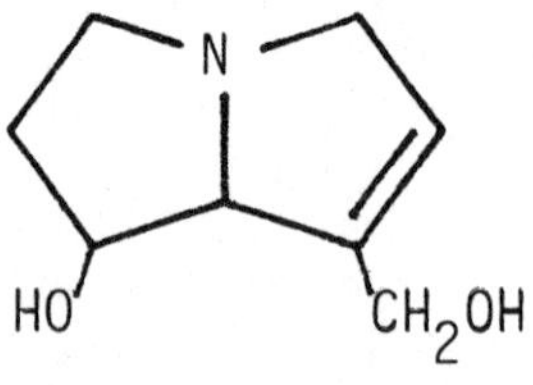

retronecine

The structure of retronecine was proved by an elegant synthesis of retronecanone, a degradation product, by N. J. Leonard in 1943.[30] The paper describing the synthesis was written mainly on a Saturday afternoon and evening, during one of Adams's hurried visits to Urbana from Washington. When Adams found that Leonard had completed the synthesis, he asked the latter to put the experimental work together that afternoon and come over to his house that evening. Adams and Leonard munched on a bag of popcorn while they worked through the evening in Adams's study, writing the paper completely enough so that it required only a limited amount of work to prepare it for publication.[31]

In addition to these varied natural-product problems and the hindered rotation series, during the 1930s Adams studied in some detail the stereochemistry of deuterium compounds. The discovery of heavy water (deuterium oxide) by H. C. Urey at Columbia in 1931 led stereochemists to wonder if the difference between hydrogen and deuterium was great enough to make an asymmetric compound with measurable optical activity. In its simplest form the question concerned the existence, in optically active form, of a structure of the type shown below. Many

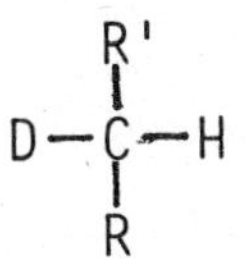

$$D-\overset{R'}{\underset{R}{C}}-H$$

chemists investigated this problem. Adams with McLean and Leffler looked for evidence of asymmetry in dideuterosuccinic acid, HOOCCHDCHDCOOH, and in 2,3-dideuterocamphane, without success.[32] An ingenious approach involved the resolution of ethyl ethynylcarbinol and its catalytic reduction with deuterium; the product was optically inactive.[33] It is not completely ruled out here that

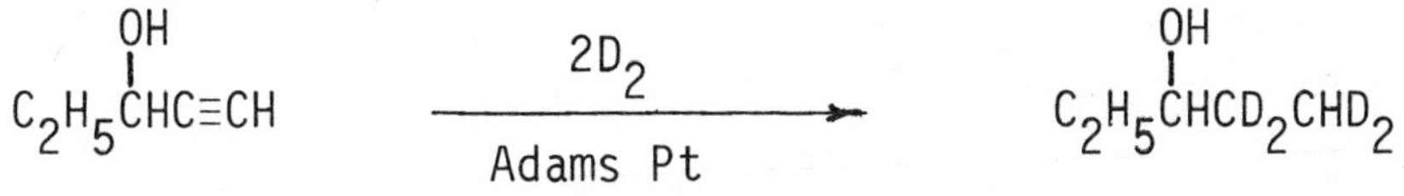

$$C_2H_5\overset{OH}{C}HC{\equiv}CH \xrightarrow[\text{Adams Pt}]{2D_2} C_2H_5\overset{OH}{C}HCD_2CHD_2$$

optically active — optically inactive

the catalyst may have caused racemization of the starting material or the intermediate allylic alcohol, $C_2H_5CHOHCD=CHD$. English workers reported a resolution of the pentadeuterophenyl derivative below, but a repetition of this work, using much purer hexadeuterobenzene, showed

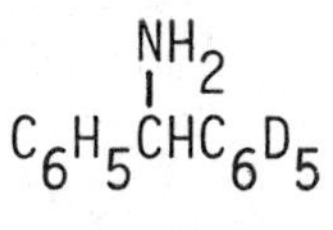

$$C_6H_5\overset{NH_2}{C}HC_6D_5$$

that the claim could not be substantiated.[34] Years later Eliot Alexander at Illinois[35] demonstrated slight optical activity caused by deuterium. Later work by others who used enzymatic or asymmetric reduction resulted in optically active primary alcohols of the type RCHDOH.[36]

During this period Adams had cursory interests in several other problems, including further work on local anesthetics. Several papers noted below deserve special mention, because they represent fields that became very active in later years and in which the Adams work was a pioneer contribution, even though not recognized by later workers.

In 1929 the polymerization of ω-hydroxydecanoic acid was found by Adams and Lycan to lead to linear polymers and to a large dimeric cyclic ester containing a ring with 22 atoms. They also studied large-ring lactones; much further work was also done by Ruzicka and by Carothers at Du Pont. Adams did not continue his research on condensation polymerization.[37]

His later paper on large-ring compounds made by the high-dilution procedure was a forerunner of work on cyclophanes by D. J. Cram at UCLA and by others.[38] Adams did not follow this up because of other problems and the intervention of the war.

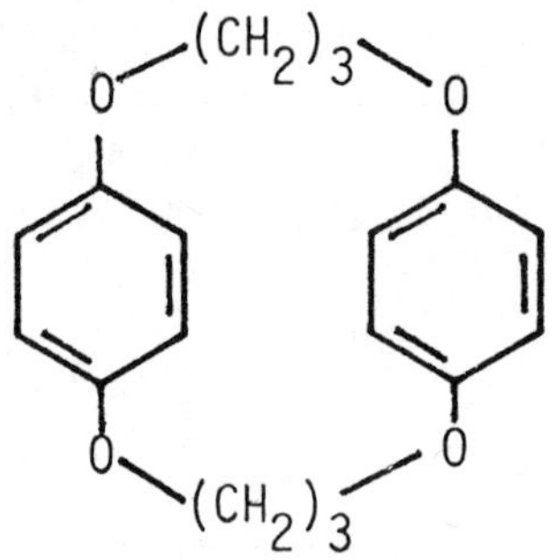

The chemistry of the carbodiimides was investigated in a single paper, with the idea that they should be asymmetric molecules like allenes and hence should be resolvable.[39] The resolution did not succeed, but further study of this little-known class by others led to their use as valuable reagents for forming phosphate ester bonds in the synthesis of nucleotides, and for the formation of peptide linkages in polypeptide syntheses by H. G. Khorana, J. C. Sheehan, R. B. Merrifield, and a host of others.

R-N=C=N-R'	R-C=C=C-R'
carbodiimides	allenes

During the 1930s there was great interest in the synthesis of polycyclic aromatic hydrocarbons to study carcinogenic behavior, particularly by L. F. Fieser at Harvard. There was also dramatic activity in the study of the sterols and sex hormones, which were reduced phenanthrene derivatives; indeed, Adams lectured on this topic before local sections of the American Chemical Society while he was president. Allene Jeanes, under Adams's guidance, made a careful study of the addition of sodium to phenanthrene, which had been reported to go in the 1,4-positions by some German workers.[40] This might have pointed to a useful way to synthesize compounds related to the steroids or sex hormones.

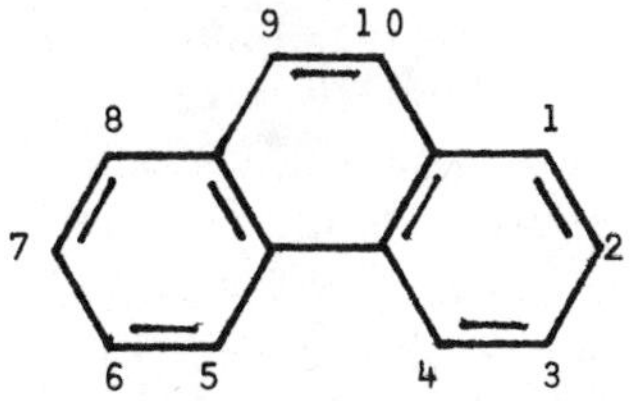

Jeanes showed that the earlier work was erroneous. Addition of sodium and other alkali metals took place in the 9,10-positions of phenanthrene, and the addition of alkali metals was greatly favored by ether solvents such as $CH_3OCH_2CH_2OCH_3$, which could chelate with sodium. N. D. Scott at Du Pont[41] first observed the favorable solvent effects of this ether and of dimethyl ether. The Jeanes paper presented two key ideas: (1) it suggested that addition of alkali metals to double bonds proceeded stepwise through addition of one sodium at a time, to form a radical anion intermediate; (2) it proposed that the promoting effect of the glycol ether was due to chelation with the sodium, as shown below. Both ideas were extremely useful in later work on reactions of organometallic compounds; the free-radical intermediate in addition of sodium to double bonds has been suggested by others.

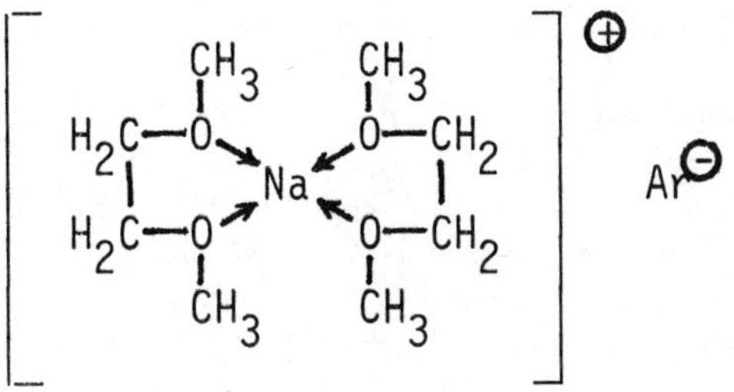

There was much chemical interest during this period in the possibility of isolating boat and chair forms of cyclohexane derivatives (the Sachse–Mohr structures). W. Hückel in Germany isolated the cis and trans forms of decalin, showing that the chair forms were stable in fused ring systems. Several workers reported isolation of stable boat and chair forms of monocyclic cyclohexanes. As a leading stereochemist, Adams repeated some of this work with R. F. Miller and found the conclusions erroneous. They then made a very careful attempt to separate boat and chair forms of a substituted cyclohexane that had no complications caused by asymmetric carbon atoms.[42] The results were negative, and the conclusion was that there was no evidence for such isomers as stable entities at room temperature. Adams did not participate in the development of conformational analysis in the following years, which produced physical evidence by 1936 to confirm the idea that boat and chair forms were easily interconvertible. The physical method of determining energy barriers to interconversion of cyclohexane conformations, using nuclear magnetic resonance, was worked out principally by Herbert S. Gutowsky at Illinois, later the director of the School of Chemical Sciences there.[43]

A survey of Adams's scientific publications that appeared in the years 1927–42, inclusive, shows a total of 176 papers; it is safe to assume that the work for the 12 papers published in 1942 was largely done in 1941. Impressive both in its quality and bulk, this body of work is doubly impressive when one considers the amount and variety of other responsibilities Adams carried during these years. Whatever the pressures of time and demands on his energy he experienced, his research group never felt that it was working under pressure. The research pace was never leisurely, but it was also never feverish. The development of the individual continued to be the most important consideration in Adams's relation to his students.

Literature Cited

1. This improvement in the quality of American organic chemistry from 1920 to 1940, as compared to earlier periods, is apparent from a careful examination of scientific journals, particularly *J. Am. Chem. Soc.* and *J. Biol. Chem* (DST and ATT, unpublished studies).
2. RA and W. M. Stanley, *J. Am. Chem Soc.*, *54*, 1548 (1932). Stanley published eleven papers from Illinois.
3. G. H. Christie and J. Kenner, *J. Chem. Soc.*, *121*, 614 (1922).
4. RA and W. M. Stanley, *J. Am. Chem. Soc.*, *52*, 1200 (1930).
5. RA and H. C. Yuan, *Chem. Rev.*, *12*, 261 (1933); also R. L. Shriner and RA, in *Organic Chemistry*, H. Gilman, ed., Wiley, New York, 1938, pp. 259–303.
6. RA and J. F. Hyde, *J. Am. Chem. Soc.*, *50*, 2499 (1928).
7. RA and R. M. Joyce, Jr., ibid., *60*, 1491 (1938) and earlier papers.
8. RA and Stanley, ibid., *52*, 4471 (1930).
9. RA and N. E. Searle, ibid., *55*, 1649 (1933); *56*, 2112 (1934).
10. RA and T. L. Cairns, ibid., *61*, 2179 (1939).
11. RA and M. W. Miller, ibid., *62*, 53 (1940) and later papers.
12. RA and N. Kornblum, ibid., *63*, 188 (1941).

13. X-ray: G. W. Clark and L. H. Pickett, ibid., *53*, 167 (1931). Ultraviolet absorption: Pickett, ibid., *58*, 2299 (1936); M. T. O'Shaughnessy and W. H. Rodebush, ibid., *62*, 2906 (1940); M. Calvin, *J. Org. Chem.*, *4*, 256 (1939).
14. RA and C. C. Li, *J. Am. Chem. Soc.*, *57*, 1565 (1935), measured rate constants and obtained an activation energy of about 20 kilocalories for the racemization of 2-nitro-6-carboxy-2′-alkoxybiphenyls. A few rate constants were reported in other papers.
15. G. B. Kistiakowsky and W. R. Smith, ibid., *58*, 1043 (1936); later important theoretical and experimental work was done at Chicago by F. H. Westheimer: M. Rieger and F. H. Westheimer, ibid., *72*, 19, 28 (1950) and earlier papers.
16. RA and E. C. Kleiderer, ibid., *55*, 716 (1933); RA and S. L. Chien, ibid., *56*, 1787 (1934); RA and W. E. Hanford, ibid., *57*, 1952 (1935).
17. RA, R. C. Morris, and W. E. Hanford, ibid., *57*, 951 (1935).
18. References in Adams's first paper, with K. N. Campbell and R. C. Morris, ibid., *59*, 1723 (1937).
19. Summarizing paper: RA, R. C. Morris, T. A. Geissman, D. J. Butterbaugh, and E. C. Kirkpatrick, ibid., *60*, 2193 (1938).
20. RA and B. R. Baker, ibid., *61*, 1138 (1939); *63*,535 (1941).
21. J. D. Edwards, ibid., *80*, 3798 (1958); RA, T. A. Geissman, and J. D. Edwards, *Chem. Rev.*, *60*, 555 (1960).
22. RA, T. A. Geissman, and R. C. Morris, *J. Am. Chem. Soc.*, *60*, 2927 (1939) and later papers.
23. RA, Madison Hunt, and J. H. Clark, ibid., *62*, 196 (1940), and later papers.
24. RA, D. C. Pease, C. K. Cain, B. R. Baker, J. H. Clark, Hans Wolff, and R. B. Wearn, ibid., *62*, 2245 (1940).
25. RA, S. Loewe, C. Jelinek, and H. Wolff, ibid., *63*, 1971 (1941).
26. RAA, 35, European Correspondence; A. R. Todd to RA, April 20, 1962.
27. Private communication from E. H. Volwiler.
28. RAA, 52, Marihuana Research, 1938–1941.
29. Key papers: RA, E. F. Rogers, and R. S. Long, *J. Am. Chem. Soc.*, *61*, 2822 (1939); RA and Rogers, ibid., *63*, 537 (1941); RA and K. E. Hamlin, Jr., ibid., *64*, 2597 (1942).
30. RA and N. J. Leonard, ibid., *66*, 257 (1944); *Senecio* alkaloids are reviewed by Leonard in *The Alkaloids*, Vol. 1, R. H. F. Manske and H. L. Holmes, eds., Academic Press, New York, 1950, pp. 108–64.
31. Private communication from Nelson J. Leonard. At that time Leonard was a post-doctorate; he stayed at Illinois and became a leading member of the department.
32. RA and A. McLean, *J. Am. Chem. Soc.*, *58*, 804 (1936); RA and M. T. Leffler, ibid., *58*, 1551, 1555 (1936).
33. RA and F. C. McGrew, ibid., *59*, 1497 (1937).
34. RA and D. S. Tarbell, ibid., *60*, 1260 (1938).
35. E. R. Alexander and A. G. Pinkus, ibid., *71*, 1786 (1949); E. L. Eliel, ibid., *71*, 3970 (1949).
36. F. A. Loewus, F. H. Westheimer, and B. Vennesland, ibid., *75*, 5018 (1953); A. Streitwieser, ibid., p. 5014, and later papers; Streitweiser used chemical reductions.
37. RA and W. H. Lycan, ibid., *51*, 625, 3450 (1929).
38. RA and L. N. Whitehill, ibid., *63*, 2073 (1941).
39. RA and L. J. Roll, ibid., *54*, 2494 (1932).

40. RA and Allene Jeanes, ibid., *59*, 2608 (1937). The chelated sodium ion in the paper has no accompanying anion and no unpaired electrons.
41. N. D. Scott, J. F. Walker and V. L. Hansley, ibid., *58*, 2442 (1936).
42. RA and R. F. Miller, ibid., *58*, 787 (1936).
43. H. S. Gutowsky and C. H. Holm, *J. Chem. Phys.*, *25*, 1228 (1956); A. Allerhand, Fu-Ming Chen, and H. S. Gutowsky, ibid., *42*, 3040 (1965) and other papers.

Government Service, 1940-48

Washington and the National Defense Research Committee, 1940–45

As the 1930s wore on, the international situation grew more threatening. The seizure of power by the Nazi dictatorship in Germany in 1933, the formation of the Berlin–Rome–Tokyo axis, and the purges and "show trials" in Stalinist Russia increased international tensions. The failure of the British policy of appeasement of Hitler, culminating in the Munich "settlement" of 1938, the British guarantee of Poland, and the Nazi–Soviet pact of August 1939 were the principal steps to war in September 1939. Britain and France were allied against Germany, which enjoyed the benevolent neutrality of Italy, Russia (for sixteen months), and Japan. Practically no human being was immune to the effects of these cataclysmic events, and Adams was no exception.

He was not an isolationist, but he did not become an outspoken advocate of American intervention in the European war, in contrast to James B. Conant, who made a radio address urging intervention in May 1940. Adams always admired Conant greatly for his courage in speaking his mind on public issues regardless of the popular palatability of his views.[1] The fall of France and the Dunkirk evacuation of the British Army in May and June 1940 brought the European war closer to Americans.

The National Academy of Sciences and its operating unit, the National Research Council, were the natural sources for the government to request advice on the scientific and technological problems posed to the American military establishment and the American economy by the German victories of 1940. Nevertheless, the Research Council and the parent academy showed limited capacity for decisive scientific leadership, and they were not effective because of divided counsels. The academy was obligated by its charter to furnish scientific advice to the federal government when it was requested, but the academy almost never volunteered advice without a specific request. The tradition remained strong in 1940.

The aggressive, imaginative leadership required to concentrate American scientific research on war-related problems and to create effective liaison with the armed services was furnished by a group including Conant; President Karl T. Compton of MIT; Frank B. Jewett, president of Bell Laboratories and of the National Academy of Sciences; and particularly Vannevar Bush, president of the Carnegie Institution of Washington. Chairman of the National Advisory Committee on Aeronautics (NACA), Bush was a shrewd and tough-minded Yankee, a distinguished engineer and inventor, and former vice-president of MIT. As the successful head of NACA, which sponsored important research in aeronautics and maintained close relations with the armed services, Bush was familiar with the Washington scene, where he had lived for several years in his capacity as head of the Carnegie Institution.

Bush and his group were increasingly concerned by May 1940 about the German victories in Europe and sought a practical way to mobilize the scientific talent of the country for military problems. Using a World War I act of Congress that created the Council for National Defense, Bush and Harry Hopkins, then secretary of commerce and Roosevelt's personal adviser and expediter, obtained an executive order from the president dated June 27, 1940. This established a National Defense Research Committee (NDRC) empowered to carry out research on problems related to warfare and to enter into contracts with educational, scientific, and industrial research organizations to perform such research. Bush was chairman; Conant, Jewett, and Richard C. Tolman (formerly at Illinois) were among the members.[2]

Conant was given responsibility for organizing research on chemical problems, which were broadly defined as including "Bombs, Fuels, Gases, and Chemical Problems." On June 25, 1940, even before the executive order creating NDRC had been signed, an informal meeting in Washington approved Conant's request that Roger Adams and W. K. Lewis be made his vice-chairmen of Division B, the division to study chemical problems.[3] After his participation in the chemical research program in World War I, Warren K. Lewis (1882–1975) became one of the country's leading chemical engineers and professor and department head at MIT.

The choice of Adams for a key position in organizing NDRC by Bush and Conant was obviously based on reasons beyond his chemical eminence and his long-standing friendship with Conant. The moving spirits in NDRC—Bush, Conant, K. T. Compton, and Jewett—were all connected with New England or East Coast institutions. Only Richard C. Tolman (1881–1948) represented the West Coast scientific community, where he was a professor and dean at Caltech in Pasadena, and even he had been trained at MIT. If Bush's "end-run" (his own phrase) to take charge of war research was to be successful (instead of leaving it to the National Academy and the National Research Council), and if the United

States was to become actively involved in the war and in war research, a broader national base for the NDRC was indispensable.

Adams was identified with the Midwest, which was then strongly isolationist in sentiment. Adams's Harvard origins were far removed, and his personal and scientific influence in other leading midwestern universities was very great. He was therefore the best man available to undertake the mobilization of organic chemists. Bush and Conant were far too shrewd to underestimate Adams's value to their project, entirely aside from his scientific and administrative ability.

As the activities of the NDRC and later the Office of Scientific Research and Development (OSRD, which included the NDRC) grew, practically every academic institution and industrial research laboratory in the country was drawn into the war research program. Whatever regional character the NDRC had at its inception disappeared completely. This was even more pronounced when the atomic bomb project was undertaken.

Adams moved rapidly to organize organic chemists for research on war problems. He wrote to a group of leading chemists describing the NDRC contract mechanism for research in universities and convened a group of about twenty chemists who met in the family room in the basement of the Adams home in Urbana in the summer of 1940.[4] The roster of the attendants included Gilman, Conant, F. C. Whitmore, Tenney Davis, probably W. E. Bachman, L. F. Fieser, R. C. Elderfield, J. R. Johnson, R. L. Shriner, L. I. Smith, C. D. Hurd, H. B. Hass, and others. Of these the only ones who had prior experience with explosives were Tenney Davis and R. C. Elderfield. The main topic of discussion was synthesis of possible new explosives, and "the work assigned at this meeting resulted in the synthesis of hundreds of compounds" for testing.[5]

Adams attended a meeting in Washington in July 1940 on explosives and took part in conferences at the Detroit meeting of the American Chemical Society in September 1940, where Conant received promises of wholehearted support from the Chemical Society and the National Research Council. Adams's influence and reputation were important factors in getting cooperation from both these groups.[6]

The headquarters of NDRC was first established in Bush's building, the Carnegie Institution of Washington, at 1530 P Street N.W., and in the early days the salaries of some of his staff were advanced by the Carnegie Institution. Funds for the initial research contracts were supplied by the Executive Office of President Roosevelt. The leading figures in NDRC in general continued to be paid by their parent institutions, as Adams was, and received no compensation from the government. During the fall of 1940 the organization of NDRC gradually took shape, and a large meeting of chairmen and staff of Division B took place in Conant's home, the Harvard president's house in Cambridge, on December 14 and 15, 1940.

The chemical work undertaken under Conant's and Adams's directions included the synthesis of possible new propellants and high explosives, the preparation of known and new potential war gases for study, the development of protective materials such as gas masks and protective clothing, and numerous other projects such as the generation of smokes for concealment.[7]

The first chemical contract, let by NDRC in November 1940, was for the synthesis of organic arsenic derivatives at the University of Nebraska under C. S. Hamilton, who had worked for many years in this field, as the official investigator. Other research on arsenicals was soon started under R. C. Fuson at Illinois and under G. B. Kistiakowsky and P. D. Bartlett at Harvard. The latter prepared adamsite with radioactive arsenic as a tracer for use in biological mechanism studies.[8] Adams's research career in two wars had now come full circle.

The NDRC research program was greatly accelerated by Conant's stay in Britain during February through April of 1941. He met the British scientific and political leaders, including Churchill and the king, learned something about the importance of radar in the Battle of Britain, and established an NDRC office in London for exchange of scientific information between British and American research workers.[8]

The growth of the NDRC activities made necessary the reorganization of the agency. Research on medical problems was clearly necessary, and the money required for that and for the NDRC programs could only be obtained by direct congressional appropriation. The Office of Scientific Research and Development was established by executive order of President Roosevelt on June 28, 1941, and was supported by congressional appropriations. Bush was named director of OSRD, which now comprised the NDRC and a new group, the Committee on Medical Research (CMR). Conant was named chairman of the NDRC, and Adams was made a member. Conant soon became in effect deputy to Bush as director.

By December 1941, when the attack on Pearl Harbor brought us formally into the war, the OSRD was in full stride; naturally the coming of war increased the urgency and scope of all research activities with possible military significance. It also made scientists ready to accept assignments that required them to move to new locations and eager to take on new research projects, which they might have been hesitant to do before Pearl Harbor.

During this period, from the summer of 1940 to the end of 1941, Adams was heavily involved in his own research program at Urbana, particularly with the marijuana problem, but was spending an increasing amount of time in Washington. He was still consulting, as he did throughout the war, for Abbott, Du Pont, Coca-Cola, and Staley and was on half-time leave from the university during 1940–41 and full-time leave from Illinois at full salary from September 17, 1942.[9] His salary was paid by the university throughout the war, and although the exact date when he moved to Washington to live is not known, it was probably in the fall of 1941, when he became the chemist member of NDRC.

W. C. Rose became acting head of the chemistry department, and Adams returned for short visits when he could. Although he could spare little time for academic affairs, his continued concern for education at Illinois and other universities is shown by two addresses prepared for the University Advisory Committee and for a Conference on Liberal Education at Illinois in the spring of 1944. As he wrote a former student in 1943, "After another year I feel our graduate schools will be so depleted that there will be very little opportunity for anything but more or less haphazard investigations. It has been my hope that enough students could be retained so that the continuity of scientific research in this country could be maintained throughout the war period but I doubt whether this will be the case".[9]

Mrs. Adams remained in Urbana, where she was active in civilian war work. She served as county chairman in the salvage drives to conserve metals and rubber and was awarded state certificates of recognition. Their daughter, Lucile, was away from home at Emma Willard School from 1941 to 1944 and then at Mount Holyoke College. During the war years Adams managed to get away to their summer place in Vermont for some vacation with his wife and daughter.[10]

Work in the crowded offices at 1530 P Street in Washington was endless and much travel was necessary, made even more difficult by the crowded wartime trains. The NDRC had to develop and maintain as good working relations as possible with the armed services, particularly with the Chemical Warfare Service (CWS). Unless this was done, the research results coming from NDRC contracts would not be examined and tested for use by the services. Furthermore, the developing needs of the services would not be referred to NDRC for its research contribution. Although relations with the armed services were not uniformly harmonious, Adams and others in NDRC, and many officers and civilian employees from the services, managed to maintain effective working contacts, which occasionally were on a cordial personal basis.[11] Some professors with important contracts were prima donnas who required suitable application of Adams's personal charm and prestige to make them reasonably cooperative.

The administration of Division B of NDRC moved in June 1942 to the library of Dumbarton Oaks at 1703 32nd Street N.W. This estate was bequeathed to Harvard, and Conant used it for research offices for the rest of the war. Adams's office was here until he left Washington in 1945.

The increase in the number and variety of research contracts led to a reorganization of NDRC in December 1942 in which the old Division B was divided into Divisions 8, 9, 10, and 11. Most of the research projects involving organic chemistry were in Division 9. Adams was formally in charge of the four divisions (8, 9, 10, and 11), although in fact each division was run by its own chief with considerable autonomy. As the atomic bomb project (Manhattan Project) developed in size and complexity, Bush and Conant devoted more of their time to this highly secret activity, and Adams became Conant's deputy for NDRC.

A brief summary indicates the work of each division, with the heads and general subjects named in parentheses: Division 8 (George Kistiakowsky, Ralph Connor—preparation, evaluation, and fundamental studies of propellants and high explosives); Division 9 (Walter R. Kirner—chemical warfare agents and protective chemical measures against them); Division 10 (W. Albert Noyes, Jr.—gas masks, smokes, smoke generators, oxygen generators, and insecticide dispersal);

Division 11 (E. P. Stevenson—oxygen-generating and -detecting devices, incendiaries, flamethrowers, and a variety of projects primarily of an engineering nature). Division 19 (Harris M. Chadwell) was first associated with the Office of Strategic Services (OSS) under Stanley P. Lovell and did research on materials and devices useful for sabotage purposes, particularly in the occupied countries. This division was known as the cloak and dagger boys, and its activities were kept secret from the rest of NDRC, although Adams had some share in them.[12]

Adams had many responsibilities in addition to his general oversight of Divisions 8, 9, 10, and 11. A list of his committee memberships, apparently dating from 1945, showed 22 committees;[13] of these five were marked "Personal" in Adams's writing: two were connected with the American Chemical Society and one each with the American Association for the Advancement of Science (AAAS), the Illinois State Board of Natural Resources, and the Nutrition Foundation.

In a long letter of April 18, 1945, to his Rotary friends in Urbana, Adams described his wartime activities in his own style in as much detail as war secrecy allowed. After a succinct account of the NDRC organization, he went on:[14]

> My own efforts, between meetings of various committees ... have been devoted to following the work of and advising the four divisions whose operations have been in the chemical and chemical engineering fields. Perhaps 50 percent of their problems have been in gas warfare, which includes defensive measures such as the chemicals used in the mask, impregnation of clothing, anti-gas ointments of different types, improved means of detecting gas in the field, and offensive measures such as search for new gases, testing the efficiency of gases, new inventions for disseminating gas, munition requirements for accomplishing certain objectives, micrometeorology to determine gas cloud movements. Besides these items, chemical warfare includes incendiaries and flamethrowers on which a prodigious amount of research has been done. The incendiary bombs being dropped on Japan at present evolved from work initiated and supported by the NDRC. The flamethrowers have been greatly improved and have been most effective in the South Pacific. The other fifty percent of the chemical work has been most varied. We have a large explosive division which has been very successful. In this war the tendency in this field has been not so much toward searching for new explosives as it has been in streamlining the older ones to produce greater effectiveness under certain conditions. This has been done chiefly by modifying compositions. Other problems are improved antifouling paints for ships and naval aircraft, hydraulic fluids and recoil oils which have essentially a flat viscosity curve between $-40°$ and $+120°$ (these are being used now in enormous volume), anticorrosion liners for shells, chemicals to be used in jet propulsion, new manufacturing equipment for liquid or gaseous oxygen which is lighter in weight and more mobile than that previously available. These are by no means all the problems in chemistry and chemical engineering but they give you a good cross-section.
>
> The nature of my assignments has not resulted in the extended trips and interesting experiences which many of my men have had. I have visited con-

tractors all over the country, some of them many times. I have also travelled to the proving grounds, many of which are unique—one in central Utah on the desert 50 miles from any habitation, one in the barren wastes of western Canada, another in Florida where tests under semi-tropical conditions can be made, and a fourth on a tropical jungle island not inhabited since 1875 and located off the southern end of the Panama Canal. While at Panama, I spent a day at Barro Colorado, the island formed in Gatun Lake when the Canal was originally filled. The island was native jungle and it is now a haven for naturalists. The United States Government owns it and a few Spanish Indians under the direction of a Department of Agriculture representative take care of the few primitive structures and of any naturalists who go there to study ... I saw wild turkeys, parrots, paraquats, toucans, anteaters, ants an inch long, wild boars, and flocks of both black and brown monkeys, just to mention some of the common birds and animals there. My memory of that trip was intense for some time, for after picking off ticks for three days, I itched jigger bites for three weeks....

The island Barro Colorado, of which Adams gave such an unflattering description, is famous in the annals of ornithology as a field station. The ornithologists used it in the winter, however, and Adams was there in late spring of 1944, when insect pests were worse than they were in winter.

Adams visited Britain in 1943 principally to see how the British war research was organized and how it was connected to the military. He also had discussions with American military men in Britain, visited British war research stations, and discussed many of the NDRC leading problems with those engaged in similar work in Britain. A diary of this trip, which he sent to Urbana when the NDRC offices closed after the war, seems to have disappeared, but he must have been in Britain for at least a month. He reported his visit to Conant in a three-page letter, accompanied by a thirteen-page memorandum titled "Notes on the Organization of Chemical Warfare in England."[15] The memo gives an incisive description of the British organization and implies, although it does not say so explicitly, that the American organization could be improved by adopting some of the British features. Adams was far too realistic to think that any fundamental reorganization involving civilian and service research groups could be carried out in the middle of a desperate war. His Rotarian letter gives further details:[14]

... As a by-product of the European trip, I renewed several old acquaintances and met many distinguished English scientists whom I had not known before. I observed London and many other cities in England which had been bombed and appreciated more fully what we in this country missed and what the Germans are now taking. Airplane travel to Europe is so simple that it is easy to visualize what may be expected in the postwar period. Two days in Ireland through the courtesy of the weather gave me an idea of that country in war-time and also allowed a visit to the National Horse Show at Limerick and a glance at many Irish people from the lowliest to the earls.

Adams was associated with the National Inventors Council, set up in the Commerce Department with Charles F. Kettering of General Motors as chairman. It screened all suggestions from the public or military of items that might be useful in the war. Of some 180,000 suggestions, seventy-seven were financed for further study, and several were used in combat.[14] He described other field trips:[14]

> A trip a year or more ago to one of the new T.V.A. projects was interesting. A valley was about to be flooded so we bought the houses for study of incendiaries, also a railroad, locomotive, cars, a tunnel and bridge to experiment on a large scale with devices for derailment, train demolition, blocking a tunnel and timed explosive charges of many kinds....

In Florida:[14]

> [I saw] all the devices, new and old, for demolition of beach obstructions and a fleet of 150 ships landing troops just as would be done on an island in the Pacific. With the exception of the firing of guns or rockets from ships, everything else was carried out as in actual combat. Confidential movies of landings and advances on the Japanese gave too intimate a picture of what happens to the boys.

A quite different aspect of NDRC research was insect and rodent control:[14]

> It pertains chiefly to DDT and its dissemination and to the search for new insect and mite repellents. Incidentally, scrub typhus which has caused plenty of trouble in the Pacific is carried by mites, just another term for ordinary jiggers. The Department of Agriculture has a station at Orlando, Fla. which I visited a few weeks ago. We support them to carry out the screening of new insecticides and repellents. Among other things, they raise 40,000 mosquitoes a day—not so easy to do—and 5,000 houseflies and bedbugs. Their pride and joy is a colony of 40,000 body lice which they keep constantly at that size, although sometimes they use several thousand a day in experimentation.

A graphic description of the growing of lice on human hosts followed.

His wish to return to his university work is clear from the last paragraph:[14]

> Occasionally I do get home for a few days but am so bogged down with accumulated items that I haven't been to Rotary. In between I am kept in touch with the gossip around the campus in one way or another. Enough said! I'm looking forward to the day when I can get back to Urbana again and settle down into efforts that really attract me.

This account does not mention his connection with the penicillin and antimalarial projects, on which he was an adviser to the Committee on Medical Research programs. Another of Adams's wartime activities, which is not given in the list of committees, was his association with the Baruch Rubber Committee. President Roosevelt appointed this committee in August 1942, with Bernard Baruch as chairman, to investigate the progress in the synthetic rubber program, about which widespread doubts existed and whose success was essential for national survival. Conant and K. T. Compton were the technical members on the committee, and Conant asked Adams and four chemical engineers to serve as consultants. How much time Adams put on the survey and in writing the report of the committee is not known; if he followed his usual practice, he would not have been associated with the survey without taking an active part in it.[16]

An informal view of Adams and the life in Washington is provided by Ralph Connor, who had taken Adams's course in stereochemistry at Illinois as an undergraduate and worked on explosives throughout the war, becoming chief of Division 8.[17] He wrote:[18]

> When I went to Washington in 1941, the members of NDRC were on a part time basis and so was Roger who was at that time Chairman of Division B, the Chemical Division (later split into Divisions 8, 9, 10, and 11). They had monthly meetings—a tempo that was not satisfactory after Pearl Harbor.
>
> I do not think Roger ever moved to Washington; that is, Mrs. Adams did not come there and he had no permanent apartment.... I often was in his hotel room. Usually he stayed at the Lee-Sheraton Hotel.... I usually stayed there as did many other NDRC people.
>
> His travel that I know about was generally near Washington—contacts with Edgewood and nearby military establishments where he had to see military personnel. Much to do at the Pentagon and Navy Dept.—It was a tiring life, even with such short trips.
>
> ... I think he travelled back and forth to Urbana throughout the war. I don't see how he stood it—but I do remember being with him in the evenings after he returned; he was full of things to talk about that related to his research progress....

Occasional poker games provided some relaxation.

> If Roger liked anything better than talking chemistry, it was winning money from someone—not that he cared about the money but it gave him an opportunity to kid someone, so every couple of months, when the right people happened to be in Washington, there would be a meeting of the "Board of Economic Redistribution." (From my standpoint it should have been Uneconomic Maldistribution.) ... The stakes were low—no one lost much money. For me, it was about as expensive as going to a movie and much more entertaining. It was dealer's choice and we played kinds of poker that I have never heard of since then. But to do this with a congenial group was, I now realize, a kind of therapy for him.
>
> There came a time when, if we were in town at the same time, we would have dinner together. Nothing fancy—a convenient hotel. We were generally too

tired to go to much trouble. We starting matching to see who would pay for the meal. It was incredible, but I won for seventeen consecutive times before I lost! Then we alternated for a few times and I started another streak. I figure that I about recovered what I contributed to him in poker. But I sure had fun kidding him! I would go into his outer office and say to his secretary, in a loud voice, "Is my meal ticket in town?" He would come out spluttering and want to match for a meal. I would protest that I felt badly about beating him so often and insist that I would buy. Of course, this made him more determined than ever to match. And when he finally did win, I cut out some of his pleasure by rejoicing that at last I could return some of his generosity.

Some time after the war, Roger got an honorary degree from Penn and I was invited to the reception. When I came to Roger in the receiving line, the first thing he said was, "I'll match you for a dollar." I won his silver dollar.

A striking vignette of Adams is given by Wendell M. Latimer, a physical chemist, student of G. N. Lewis at Berkeley, and lifelong faculty member there:[19]

Your request for anecdotes on Rodger (sic) Adams is indeed difficult to answer. I have many vivid memories of sessions in his room at the Lee Sheraton in Washington and at Dugway Proving Grounds. High-low six card stud is one of his favorite games. I recall Rodger as the "life" of the party in Panama, dancing with the general's secretary. However these gay moods are outnumbered by memories of Rodger seriously discussing some scientific problem, preferably however with a high ball in one hand. If I were shipwrecked on a distant island, I would hope that Rodger Adams would be with me. I am sure that he would figure out some way of getting off and that we would have a whale of a good time while we were doing it.

After the invasion of Europe from Britain in June 1944, the defeat of the desperate German gamble in the Battle of the Bulge at the end of 1944, and the implacable advance of the Russian armies on Germany from the east, it became clear that the end of the war in Europe was in sight. The OSRD organization made plans to disband, and Adams wrote many letters to locate jobs for those of the OSRD staff who did not have positions to return to.[20] In this characteristic action he was, as usual, successful. He and Conant also arranged for the NDRC histories to be written; the volume on chemistry was edited by W. Albert Noyes, Jr., covering the work of Adams's NDRC divisions.

A letter of May 1, 1945, from Adams to Jewett shows that he had been pondering postwar relations with the Russians and had a constructive idea for consideration. The concern shown by Adams in this letter for the reconstruction of international relations undoubtedly led to his later mission to Germany and his two journeys to Japan to make recommendations for the reorganization of Japanese science. He wrote:[21]

For the past few weeks I have intended to talk with you about a matter which has been very much on my mind. The future of the world in general will depend to a very large extent on a mutual understanding between the Russians and ourselves. Personally, I have always had the greatest respect for the Russians, and even more since the war. Unfortunately the Hearst newspapers and various writers have instilled into the American public a certain fear of Bolshevism and of the Russian people as a whole, which I believe is not justified.

There is no better way of creating a friendlier feeling and eventually complete understanding than through scientists. I am therefore making this suggestion to you for your consideration. Would it not be possible for the National Academy to invite to this country ten or a dozen of the leading academic scientists merely to get acquainted with the scientists of this country and to see how we function in our universities? At the same time opportunity would be given them to visit some of the interesting spots in the United States and possibly some of the industries. This sort of program would require funds so that their complete expenses might be paid. I feel relatively certain that the Rockefeller Foundation would entertain a proposition for funds to support this visit, or in case they would not, there certainly are other agencies which would do so.

My idea is that if negotitions [sic] were started now, then when the Germans have completely collapsed and matters have settled down a trifle in Europe these men could be brought over, perhaps early in the fall. There would be several questions to be decided, such as whether the medical scientists would be represented as well as the physical and biological sciences. A small group such as I have suggested would be more feasible since their travel and visitation to various universities would not involve too much administration as would be the case if thirty or forty were to ride in a group. Moreover, it would be possible to have more informality in the way of smokers and entertainment than with a larger group. There is a strong probability that most of our major institutions would be glad to entertain such men at their institutions. I am sure also that the technical societies, such as the American Chemical Society and the American Physical Society would lend what aid they might in a project of this kind.

Naturally, the State Department would be involved but I can hardly imagine that there would be any objection on their part. A more important item would be the method of approach to the Russians. Colonel Faymonville is one of the very few men in this country who knows and understands the Russian people and at the same time is acquainted with their top political set-up. His advice would therefore be most helpful and useful....

His wartime experiences in Washington and elsewhere were bound to make Adams think about the postwar world, with his keen, very receptive mind and his broad interest in people (and peoples) as individuals. This letter is the first indication we have of Adams as a scientific statesman on a global scale, a role he occupied in varying degrees for the rest of his life. Henceforth he was not merely a brilliant chemist, teacher, and administrator from Urbana, Illinois; he became a world figure in science and an ambassador of American science to other scientific cultures. Adams's pragmatic and practical turn of mind was averse to elaborate philosophical generalities, and his role in international affairs was based on his perceptions of concrete actions that could be taken to improve a given situation.

The winding down of the OSRD organization in the summer of 1945 forced Adams to think of his own future and led to much correspondence and consultation with President Willard and the chemistry department at Urbana. The department had recommended on April 26, 1945, and the deans had approved, that Adams be appointed a Distinguished Service Professor at an increase in salary of $2,000, to $12,000.[22] Although the salary increase was granted, the title was not, presumably because such a category did not exist at Illinois and the president or board of trustees did not wish to establish it.

Adams wrote President Willard on August 28, 1945, saying he wished to return to Urbana on October 1, 1945. He also wished to give up his course in stereochemistry, which he first gave in 1920, and did not wish to resume the headship of the department for a semester, having W. C. Rose continue as acting head. Willard and Rose agreed that the latter arrangement would be unsatisfactory, because it would place both Rose and Adams in a very difficult position. Adams persisted on September 17, 1945, in his request to have a respite of a semester from the headship; he said that his physical energy was not what it had been when his war work started in June 1940. The grinding and frequently frustrating work in Washington, plus his other responsibilities in those years, had taken a toll of his seemingly boundless stamina. Furthermore, Adams was now fifty-six and he had worked extremely hard since his undergraduate days at Harvard.

His plans to return to Illinois in the fall of 1945 were prevented by the pressing request of Bush and Jewett that he spend several months in Germany as technical advisor to General Clay, head of the American occupation forces (OMGUS, Office of the Military Governor, Germany, U.S.). When Adams did return to Urbana in March 1946, it was as head of the department.

As in most of his career, Adams never wrote a connected account of his experiences and his conclusions from his Washington years; his letter to the Urbana Rotary Club is the nearest thing to it. He was fulfilling his obligations as a citizen; he probably occasionally enjoyed being in a position of authority where he knew what was happening in many secret and some fascinating projects. Scattered references in his letters show his frustration and occasional anger at dealing with the government and military bureaucracy.

His wartime services were recognized officially by award of the Order of the British Empire, signed by King George VI, and by the U.S. Medal for Merit, signed by President Truman. His war service undoubtedly played a part in many other awards and honorary degrees he received. On the whole, he probably took satisfaction from his performance in a most complicated and stressful position.

Literature Cited

1. Conant, pp. 211 ff; on Conant's courage, RA to E. J. Corey, March 24, 1971; RAA, 36, C.
2. The official history of the Office of Scientific Research and development (OSRD), of which the NDRC was a unit, is in J. P. Baxter, *Scientists Against Time*, Little, Brown, Boston, 1946; the order establishing the NDRC is Appendix A. *Pieces of the Action*, by V. Bush, Morrow, New York, 1970, gives other aspects of the story and the flavor of Bush's personality; the institutional history by Irvin Stewart, *Organizing Scientific Research for War*, Little, Brown, Boston, 1948, gives the administrative history and has numerous references to Adams's activities in the background of the OSRD enterprise.
3. Conant, p. 242; *Science in World War II: Chemistry*, W. A. Noyes, Jr., ed., Little, Brown, Boston, 1948, p. 4. This is the official history of the chemistry components of the NDRC for 1940–46, written by several authors and with a foreword by Conant and Adams; cited hereafter as Noyes.
4. RA to Henry Gilman, July 15, 1940; presumably a similar letter went to the other members. A note by Professor Gilman on this letter says that the meeting in Adams's home was held "shortly after" and names a few of the participants; also on this meeting, Noyes, p. 35.
5. Noyes, loc. cit., and pp. 52–7.
6. Noyes, pp. 5, 155; Conant, pp. 242–3.
7. Noyes, pp. 5–8; pp. 157 ff.
8. Conant, pp. 248–71.
9. Memorandum and letter, RA to E. L. Bowles, October 4, 1945; National Archives, OSRD (R. G. 227), NDRC, Roger Adams file, folder August 1944–July 1945. His leave from Illinois is documented in the Illinois Archives, Willard papers, 2/5/15 President's File, 364; it was approved by President Willard October 16, 1942; Adams's speeches: RAA, 54, Speeches 40–46; on graduate research: RA to DST, March 19, 1943.
10. Information from Lucile Adams Brink. The certificates are in RAA, 8, 1943–1944.
11. On relations of NDRC and the armed services, see Noyes, pp. 141 ff; Bush, *Pieces of the Action*, pp. 75 ff. A long memorandum by Conant and Adams to Bush of January 6, 1945, reviewing relations between NDRC and CWS and indicating that marked improvement had taken place partly because of the formation of a CWS–NDRC committee to facilitate direct communications by bringing together representatives of NDRC and CWS is in National Archives, OSRD (R.G. 227), NDRC, Roger Adams file, folder Aug. 1944–July 1945. This memo pays special tribute to W. Albert Noyes, Jr., for effective work in bringing about better cooperation with CWS: "We cannot praise too highly the work of Dr. Noyes."
12. Noyes, pp. 11–14; IX–X; and passim. An entertaining account of some Division 19 activities is in S. P. Lovell, *Of Spies and Strategems*, Prentice-Hall, Englewood Cliffs, N. J., 1963.
13. "Miscellaneous Committees of which Dr. Adams is a member," no date, National Archives, OSRD (R. G. 227), NDRC, Roger Adams file, folder Committee Memberships.
14. RAA, 7, Rotarians; RA to Rotarians.
15. Letter, RA to Conant, October 8, 1943, with memorandum, National Archives, OSRD (R. G. 227), Central Classified Correspondence, Chemical Warfare (now declassified).
16. Conant, pp. 305–28; RAA contains a document signed by Conant, which states that

Adams is a consultant to the committee, with Adams's signature for identification. RA is not mentioned in F. A. Howard, *Buna Rubber*, Van Nostrand, New York, 1947, but few of the members of the Baruch committee are. A list of the documents deposited in the National Archives by RA mentions papers from the Baruch committee.

17. Stewart, op. cit., p. 88.
18. Ralph Connor to DST, December 4, 1978.
19. W. M. Latimer to N. A. Parkinson (of *Chemical Engineering News*), October 20, 1949. The last two sentences of this letter have been quoted in sketches of Adams and attributed to a "student" of his, which Latimer never was.
20. Letters in National Archives, OSRD (R.G. 227), NDRC, Roger Adams file, box 1554, folder Aug. 1944–July 1945.
21. May 1, 1945, RA to F. B. Jewett (in his capacity as president of the National Academy of Sciences); National Archives, OSRD (R. G. 227), NDRC, Roger Adams file, folder Aug. 1944–July 1945.
22. Illinois Archives, 2/5/15 President's File, 364.

Germany, 1945–46

Adams was in Germany from November 1945 to February 1946 as scientific advisor to General Lucius D. Clay, deputy military governor to the Office of Military Government for Germany U.S. (OMGUS). In 1947 and again in 1948 he was to lead delegations of American scientists to Japan to make recommendations to General Douglas MacArthur about the democratization of Japanese science, which will be discussed later. These three missions represented a major expenditure of time and energy in the national service by Adams, and the results were significant in the development of American postwar relations. Coming after his arduous duty with NDRC and at a time when his primary desire was to return to his research and teaching at Urbana, these undertakings demonstrated anew Adams's unselfish devotion to the national interest, comparable to that of the John Adams–John Quincy Adams branch of his family. They also showed him utilizing his administrative skill and personal qualities to the fullest.

Few experiences in Adams's life were as moving and as sobering as the months he spent in occupied Germany in 1945–46. He had lived in Germany for a year in 1912–13, learning the language, observing the university system, and making many German friends when the country was at the height of its scientific, political, and military strength. In 1945 Germany was completely conquered militarily, its territory was divided and occupied by four different powers, its cities were largely destroyed, its economic system had collapsed, and life for all Germans was reduced to a constant struggle for the barest elements of shelter and food. Adams's whole career was based on an antiauthoritarian philosophy, and he abhorred the systematic atrocities of the Hitler regime, but he did not regard the surviving German people with a vengeful or doctrinaire attitude. His approach as usual looked toward the future, and he asked how to restore some semblance of normal life in Germany and to prevent a recurrence of the Nazi terrors. Formal reports and official correspondence tell the story of his German months, but far more vivid are the accounts in a small pocket diary and in the letters he wrote to Mrs. Adams twice a week. These give an unforgettable picture of the misery, disorganization, and virtual anarchy under the four occupying powers. We base our account mainly on these sources.

After the successful cross-channel invasion of the European continent in June 1944, the OSRD organization began to plan for its own demobilization. At the request of secretary of the navy, James Forrestal, the National Academy of Sciences (NAS) established the Research Board for National Security (RBNS) in 1944.[1] This board was to succeed the OSRD, to complete its unfinished business, and to continue to provide scientific advice to the armed services. Adams was a member until he resigned to go to Germany in 1945. The RBNS was later abolished by President Truman when legislation was pending for what became the National Science Foundation. The RBNS was at best a stopgap group and seems to have accomplished little.

The NDRC discussed the fate of German science after the war on numerous occasions in connection with postsurrender plans for German disarmament. The CWS–NDRC Committee on Jaunary 10, 1945, for example, considered reports by the biochemist Joseph Needham on world science and one by Leo T. Crowley on German economic and industrial disarmament. Adams was chairman of a committee of NAS under OSRD auspices that studied the policy toward German scientific research and engineering and issued a long report.[2] In addition, the NDRC was actively involved in sending technical intelligence teams to Germany and the occupied countries, partly to determine what had been done on the making of an atomic bomb (the "Alsos" mission).[3] The survey of the other accomplishments of German science and technology during the war, the FIAT missions (Field Information Agency, Technical), was under army auspices.[4]

The Potsdam Protocol of July 1945, signed by the United States, Britain, Russia, and France, provided, among other things, for the elimination of German war potential, the industrial disarmament of Germany, the payment of reparation to countries that had suffered from German aggression, and the establishment of a self-supporting Germany.[5] These aims all required that the military governments in the four zones of Germany should have a large amount of technical advice. How does one select the "war chemicals," for example, whose manufacture should be prohibited in Germany? Adams was almost uniquely qualified to act as scientific adviser to General Clay by his scientific eminence, his experience in Germany as a student, his very broad knowledge of the chemical industry, and his concern about postwar international relations.

Although Adams hoped to return to Urbana in the fall of 1945 (without acting as head of the chemistry department), the request from the government that he spend six months in Germany introduced a new factor. In a handwritten letter of September 9, 1945, Adams asked for a personal conference with Illinois President Willard. After this and a meeting with the chemistry department staff, Adams reduced his request to a four-month stay in Germany instead of six months. The secretary of war wired President Willard on October 16, 1945, asking for Adams for four months on a very important assignment for which he was the outstanding candidate. The trustees at Illinois approved Adams's leave for four months on October 18.[6]

Adams cleared his consulting connections, Du Pont, Abbott, Staley, and Coca-Cola, with the War Department through E. L. Bowles, the expert consultant to the secretary of war, so that there would be no possibility for conflict-of-interest claims, and also made sure that he would be on the same administrative level as the political and legal advisers to General Clay. Adams said he was dependent in part on his consulting fees to carry his fixed obligations.

Adams left Washington for the military base at Frankfurt on November 6, 1945,[7] flying from Bermuda to Paris via the Azores in 24½ hours. Held over by weather, he saw great numbers of Americans, both military and civilian, in Paris but few French and few children. Food, soap, and cleaning materials were scarce,

Illinois University Archives

Mrs. Adams, Lucile and Roger (1945).

Illinois University Archives

Adams receiving the Davy Medal of the Royal Society from General Lucius D. Clay, Berlin, 1945.

and prices were five times American prices. Arriving in Frankfurt on November 11, he was taken to the nearby city of Höchst where his committee was staying, and here he was billeted in a cold but undamaged house. Two members of FIAT, Colonel Osborne, a former member of the NDRC, and H. P. Robertson (1903–61), a distinguished mathematician from Princeton, saw him there, and from them he got an idea of the "stupendous complications" involved in getting scientific information and formulating policy from it. The difficulties in FIAT were caused primarily by lack of staff and constant personnel changes.

A quick trip to Heidelberg gave him a chance to visit the distinguished organic chemist Karl Freudenberg, who had been the Adams's guest in Urbana, and to discuss German scientific publications. A denazification court had already cleared Freudenberg and allowed Heidelberg University to start teaching its curricula in agriculture and medicine.

The Quadripartite Committee on Restriction of German War Chemicals (and research on them) with its Russian, British, French, and American members was stationed in Berlin. Adams's trip there was made by overnight train to Helmstedt at the border of the Russian zone, where he and others were "herded into closed trucks for Berlin, arriving at 12:45 (pm)." He had an interview with General Clay and got the impression that Clay was not clear about a scientific adviser's duties. Clay asked Adams to spend a week interviewing his staff members and to "submit to him a recommendation about what he (Adams) should do." Adams wrote his wife that the job "looks complicated" due to the arrangement of assignments in different organizations; there were several thousand staff members.

At this time the Allied military governments in Germany were proceeding on the assumption that quadripartite control would be made to work, although all indications were that it could not. American occupation policy either was not formulated or, when it was, was unrealistic and unworkable; Clay was an able administrator and politician and was obliged to develop American policy largely on his own.[8] Adams constantly refers to the defects in American policy and the generally low quality of American personnel, both military and civilian, in the middle and lower ranks, though he came to have high opinion of Clay, with whom he remained on friendly terms the rest of his life.

In his letters home Adams noted the destruction in Berlin, the black market, the ruins of the Reichschancery, where Hitler had died in his bunker, and the cross-zone kidnapping of Germans for information by the Russians. He found that he was to be responsible for formulating recommendations for the control of research for the Committee for the Liquidation of War Potential (CLWP). He expected to see General Clay again and planned to return home if his position was not clarified satisfactorily. After another interview with Clay, Adams wrote again, "I think Clay doesn't know exactly what to do with me. It is a bit amusing and a game which I am enjoying. I've already submitted a memorandum to Clay with certain recommendations, and it's up to him to take the next steps."

Adams had a good Thanksgiving dinner with a FIAT group, bought some

German stamps, and, after attending a quadripartite committee meeting that got nowhere, was unable to decide what authority he had, where he was expected to report, and where the center of power was located. He had some good discussions privately with the Russians. He looked into the situation of German scientific publications, which was very complicated because a given publication might have its printer, editors, and offices in three different occupied zones of the country. Although he wanted to revive the German publications, progress was "discouragingly slow." Amid these various frustrations great news broke: within the space of three days Adams heard that he had been awarded two major medals, the Davy Medal of the Royal Society of London and the T. W. Richards Medal of the Northeastern Section of the American Chemical Society. In a few days the Davy Medal arrived in Berlin, brought from England by Robertson, and General Clay made the presentation, reading the citation from the Royal Society.

By November 30 Adams had as personal staff George Mercer, of the London OSRD office, and Lieutenant Karl Olson, U.S. Navy, who had been a graduate student in political science at North Dakota and at Yale. Adams paid high tribute to them both. Olson wrote a brilliant and moving report on the general postwar conditions in Germany, one of the most interesting documents to come out of the Adams mission, as well as reports on German scientists, on the renowned German compendia on organic chemistry *(Beilstein)*, on inorganic chemistry *(Gmelin)*, and on the chemical journal *Berichte*.[9,10] Part of the *Beilstein* office and library had been taken by the Russians, but Adams talked several times with F. Richter, the *Beilstein* editor. Adams concluded in his report that resumption of publication was not possible under the prevailing conditions; Olson's report suggested that the compilation and publication of *Beilstein* might be continued at Leverkusen in the British zone.

The reopening of universities and research institutions involved the problem of non-Nazi German scientists in U.S. detention camps;[9] many were under arrest merely because of their high civil service ratings. The status of the Kaiser Wilhelm Institutes (later renamed the Max Planck Institutes), once the leading research centers in Germany, was uncertain because under the American policy (contrary to the other three powers) the former directors were mandated for arrest and eventual denazification procedures. Adams's view that the American denazification procedures were more inflexible than those of the French and British and were inherently unworkable is supported by later studies.[11]

On November 27 Adams wrote that he liked General Draper, head of the economic directorate, but he found his fellow committee members on the study of German research regulation inexperienced and of limited compentence. "It means, however, that I'll have more influence.... It's interesting because I've got the ace card when it comes to a showdown—it's just playing politics with high army generals where actually they have a little worry about a scientist, strange to say." Adams's years as an expert poker player gave him a relaxed view of the general situation. One must sympathize somewhat with the bewilderment of a professional military man in dealing with Adams, who was anything but a professor from an ivory tower.

By December 5 General Clay's chief of staff had accepted Adams's proposal to set up the research committee at a staff level, and Adams was appointed scientific adviser to Clay, assigned to the economic division and reporting directly to General Draper, a move that gave Adams the rank he needed to be effective. He was busy making study outlines of methods for control of German research. The "Liquidation of War Potential" (LWP) was "an awful mess" with continual changes in plans; he hoped to meet with the British representatives to get agreement with them. The French members gave a long dinner and a great party, and after a British dinner Adams taught them all to play poker, but "the playing was not wildly exciting."

Adams's own billet was a large house (with a huge bathroom 20 feet × 20 feet) near the Wansee Lake in Berlin, but it had no heat and it was so bitterly cold that he did not move in. A bright spot was that in December Adams had his turn as chairman of the technical committee of the Quadripartite Commission, so he was doing what he came for.

On December 19 Adams was discouraged with the Quadripartite Commission; he had three documents ready to finish and he had spent hours trying to get the Russian members to agree to meet. He planned to visit the U.S. zone at Christmas to look over some research establishments; he had constant requests from people in the United States to try to locate relatives or friends in Germany, which he handled as conscientiously as possible. Finally on December 21 the Russian delegates came; the three documents were agreed on and were submitted to higher authority. One of these documents was his plan for control of permitted German research, and it passed with minor changes, so that there was now agreement on policy, control, and assignment of responsibility to the economic division. "I feel I have accomplished a lot assuming the Big Boys [the Quadripartite Allied Control Council] don't reject it—and I hardly see how they can—we have reached the major goal I hoped to attain while here." With this achievement he felt that he could afford the time to travel.

He attended a good Christmas party at Frankfurt, where he found his colleague N. J. Leonard doing an excellent job of photostating and shipping to the United States the research reports of the I. G. Farben, the huge German chemical cartel. On December 30 he described an auto trip to Munich; he was impressed by the Autobahns (the superhighways built by Hitler), though there were many destroyed bridges. He visited several chemical plants and talked to some Germans. Administration was difficult in the U.S. zone, partly because of rivalry between the U.S. Third and Seventh Armies. Back at Frankfurt for the New Year, Adams arranged an "egg-nog" party, though eggs were not available. Leonard, Adams's guest for the holiday, spent much time stirring peanut butter and dried milk into liquor to give it the proper consistency. The "egg-nog" was universally approved and consumed in quantity, and the party raised Adams's stock considerably with his associates.[12]

From Frankfurt he visited the detention center coded "Dustbin" by the British (a term synonymous with the American "ashcan" or "garbage can").[12,13]

> [Dustbin] is an old castle located about thirty miles north of Hoechst outside of a tiny village. It was the headquarters for the German army during the Ardennes drive. It is now being used for housing various Germans who occupy important positions and who are now retained for interrogation. There are several IG directors and other prominent politicians and technical men there. They are required to clean their own rooms, make their own beds and wash their clothes. The building is heated and they were receiving adequate food last winter. The grounds are spacious and they are allowed to roam freely around. So far as I know, there were no third degree tactics used in attempting to get information. For the most part, any of these men were quite willing to answer questions and disclose any information they had.

Adams also talked to Reppe, the German industrial chemist who had led a brilliant scientific and industrial development of the chemistry of acetylene to replace the natural fats and petroleum products that Germany lacked in sufficient amounts. Not surprisingly, he found Reppe a "prima donna."

After a car trip to Göttingen in the British zone, Adams talked with Telschow, the acting director of the Kaiser Wilhelm Gesellschaft (which ran the research institutes), and also with the distinguished organic chemists from Göttingen University, A. Windaus and H. Brockmann. In the "interesting and fruitful 12 days" in the U.S. and British zones, Adams "accomplished one-half" of what he hoped, but he felt that many people who would be good future leaders in Germany were being antagonized.

On January 12 he contrasted the Russian denazification policy with the American: the Russians put the high Nazis to chipping bricks but allowed the technical people to run the plants to teach the Russians, who took the chemicals and dealt with the technical people later. The American policy dislocated everyone, and the plants could not be operated. At least the Quadripartite CLWP had accepted two of the reports from Adams's group; they left two for its next meeting.

By January 20 Adams was well satisfied with the completed report on detailed control of research in the U.S. zone. The report on the revival of German scientific publications was to be ready in a few days. By February 3 the Coordination Commission accepted the report on research control, and Adams worked with the legal department to draft laws. He also sent General Clay a memo on the plight of non-Nazi scientists for the record. When the legal drafting was done, there was still difficulty in getting satisfactory French and Russian translations, but on February 6 Adams delivered two papers to General Draper for his signature: "Research Policy and Control" and "Mechanism of Control of Research in Germany in the U.S. Zone." Draper complimented him: three months was an absolute record for getting an important paper through on a quadripartite basis.

The remainder of Adams's stay was devoted to a "wonderful" visit to

Potsdam, to plans to return to the United States, though Draper urged him to stay, and to conferences with technical officers, Clay, and the American press. Adams felt that the British were the most practical of the four powers; the French goal was to block everything so that the Americans would have to feed the Germans. The Russians were reasonable in all but a few things, but there was no hope of real decisions with the four-zone policy.[7,14] These contemporary comments by as shrewd and experienced an observer as Adams are of considerable historical importance in the development of postwar Germany.

There were some lighter sides to his experiences in the grim, wintry chaos of occupied Berlin. On one occasion American junior officers seized two riding horses belonging to Otto Warburg, a Nobel Prize–winning chemist and biochemist. Unable to obtain their release, he appealed to Adams. The elderly scientist told of steadfastly refusing Russian blandishments of a scientific directorship in their zone; he preferred to stay with the Americans. After asking lower officers in vain, Adams succeeded in getting two horses sent to Warburg by direct order from General Clay's office.[15] Adams wrote 20 years later: "General Clay is a very clever individual and very versatile. Actually I never became well acquainted with him while in Germany because I chose not to attend his staff meetings, which were strictly along military lines. When there was something essential for me to have done, he had a WAC captain who could get things through more quickly than I could through General Clay; so instead of individual conferences with him, I had them with his secretary."[16]

When Adams was due to return to this country, he found that his travel priority was no longer valid because of a change in regulations. His reaction to this, although not documented in detail, can be inferred from another case; as he wrote Mrs. Adams, he was tired, and this was the last straw. His blunt, angry protests eventually got him transport home, and he resumed his place at Urbana around March 1, 1946.

Adams accomplished about as much as could be expected, considering the extraordinary complications of the military governments. He did something to improve the lot of non-Nazi scientists, he eased the revival of the German compendia, *Beilstein*[17] and *Gmelin*, and of the German scientific periodicals, and he demonstrated to OMGUS the importance of the technical problems that confronted them in dealing with German industry. No one of comparable scientific standing succeeded Adams, and soon the whole policy of occupation was changed, with the Russians withdrawing from the Allied Control Council and openly ruling their zone as they saw fit.[18]

Adams felt strongly that the Western powers were acting wrongly in seizing all the assets they could, including research results of German industry. He wrote Jewett that the question was "whether the fact that we won the war justifies our acquiring [all the] government as well as private information we can get our hands on."[13] This view, which he expressed more than once, was based on his strong belief in the rights of private property, but, more broadly, on the recognition, which eventually became general, that if the West Germans were ever to become self-supporting with a reasonable standard of living, they must be allowed to retain sufficient means of production to rebuild their economy. This is another example of Adams's instinctive grasp of economic realities.

Returning to Urbana, Adams was able to revive his own research program and to resume leadership of the chemistry department in the trying postwar period. His performance in Germany, however, ensured that he would be called on for other overseas missions.

LITERATURE CITED

1. NAS Archives; NAS: Research Board for National Security, 1944; the need for this group was described by Rear Admiral J. A. Furer, *Science, 100*, 461 (1944).
2. National Archives, OSRD (R.G. 227), Roger Adams, box 1556, minutes of CWS–NDRC Committee; a copy of this long report, *The Treatment of German Scientific Research and Engineering from the Standpoint of International Security*, by the Technical Disarmament Committee Project 3, RA, chairman, July 2, 1945, a study by an NAS committee, under auspices of OSRD and NACA; is in RAA, 58.
3. National Archives, as cited, box 1554, Scientific Intelligence Mission, Office of Field Services; ibid., A. T. Waterman to J. B. Conant, August 21, 1944, cc. to Adams; Irvin Stewart, *Organizing Scientific Research for War*, Little, Brown, Boston, 1948, p. 134; S. R. Goudsmit, *Alsos*, Schuman, New York, 1947.
4. The FIAT missions, which found, among other things, Reppe's work on the synthesis of cyclooctatetraene from acetylene and the many long-chain organic compounds resulting from it that kept the German economy going during the war, were sponsored by various branches of the U.S. Armed Services.
5. RA, *Chem. Eng. News, 24*, 2486 (1946); a more complete summary of the Potsdam protocol is in General Clay's book, Ref. 8, pp. 40–3.
6. Relevant documents are in Illinois Archives, President A. C. Willard, 2/5/15, 364, Chemistry. Because Adams's German trip was officially sponsored by NAS, some correspondence and his official report to F. J. Jewett, president of NAS, are in NAS Archives: GOVT: International Relations, Office of Military Govt. in Germany: US Scientific Adviser, Adams, R. Some correspondence with E. L. Bowles, the technical advisor to the secretary of war, is in the National Archives, as cited: Roger Adams file, Aug. 45– .
7. The manuscript diary covers the period November 6, 1945–January 23, 1946, RAA, 58, Travel Log. This is little more than an appointment book; much of it is intelligible only in connection with Adams's letters and his report to Jewett of March 5, 1946 (NAS Archives, citation in full in Ref. 6). RA letters to Mrs. Adams are in RAA, 5, Family Correspondence; Nov. 10–Feb. 24, 1946.
8. E. N. Peterson, *The American Occupation of Germany*, Wayne State University Press, Detroit, 1977; L. D. Clay, *Decision in Germany*, Doubleday, Garden City, N.Y., 1950.
9. *Report on the Activities of Roger Adams, Scientific Adviser to the Deputy Military Governor of Germany*, submitted by RA to F. J. Jewett, president of NAS, March 5, 1946; NAS Archives, citation in full in Ref. 6.

10. Attached to Adams's report, Ref. 9. Olson's name is sometimes spelled Olsen. Olson's report on postwar Germany is in RAA, 58, along with numerous other documents from Adams's German mission.
11. Peterson, op. cit., pp. 139 ff.
12. N. J. Leonard, private communication. Adams's diary runs: "Jan. 1. Late up. Mac and I prepared egg-nog. No lunch. Party 2:30–6:30. Dinner at 7:00. More egg-nogg [sic] till 10:00 P.M. Jan. 2. Late up. Morning small items. [Trip to] Dustbin in Br. [British zone]. Lost on way—finally found Cransberg at 4:00 P.M." On January 3 Adams saw Reppe, "not interested in U.S. job."
13. RA to F. J. Jewett, June 25, 1946: NAS Archives, GOVT: International Relations: Office of Military Govt. for German: General. This difference in English usage of "dustbin" has led some unwary writers into amusing errors.
14. RA to Henry Gilman, January 21, 1946.
15. Details in RAA, 58, General Memoranda, 11/14/45–3/1/46.
16. RA to Gilman, October 5, 1966.
17. Friedrich Richter, *75 Jahre Beilsteins Handbuch der Organischen Chemie,* Springer, Berlin, 1957, p. 20, paid grateful tribute to Adams's aid in reviving *Beilstein.*
18. See Clay, op cit., pp. 142–84; Peterson, op. cit., says little about the Russian separation.

Japan, 1947, 1948

The need for a group of American scientists to visit Japan and make recommendations about changes in the organization of Japanese science was early felt by General Douglas MacArthur and his staff, the Supreme Command for the Applied Powers (SCAP), the rulers of occupied Japan. The National Academy of Sciences was asked to organize the undertaking, and at least as early as October 1946, President Jewett of NAS and Adams tried unsuccessfully to persuade Ralph Connor to accept such an assignment.[1] Nothing definite followed until the summer of 1947, when Secretary of War Patterson formally requested Jewett to set up the mission.[2] Jewett asked Adams to serve as chairman; the mission was to spend about forty days in Japan, it was to have six members, its transportation would be furnished by the army, and its other expenses would be paid by a grant to NAS from the Rockefeller Foundation. Adams readily secured permission from Illinois President Stoddard and Dean McClure to make the trip, especially since it would come in the summer.[3]

The committee was a distinguished one: W. D. Coolidge, physicist of General Electric; R. W. Sorensen, electrical engineer of Caltech; W. V. Houston, physicist and president of Rice Institute; M. K. Bennett, economist of Stanford; and W. J. Robbins, botanist of the New York Botanical Garden.

The trip was strenuous. After their arrival in Tokyo on July 31, 1947, the committee held conferences with Japanese scientists and administrators and with the occupying authorities, then proceeded to about six other Japanese cities. Lack of time prevented a visit to Seoul, Korea, which had been in the original itinerary. In fact, the work in the oppressive humid summer heat of Japan was more arduous than the schedule suggests. Japanese hospitality was elaborate and time-consuming, and Adams was under pressure to prepare and give two major speeches, while all members of the group toiled to write the report. Departure from Japan was on August 29, 1947,[4] and Adams wrote Jewett on September 30, 1947,[5] "The trip to Japan was so strenuous that I felt it imperative to take a few days vacation before and after the [American Chemical Society] meeting in New York so that I would feel refreshed enough to start the fall semester."

That the mission performed an intensive study is shown by the formidable list of universities, industrial laboratories, and industries visited, as well as by the many meetings with Japanese groups and sections of the occupation government, including a conference with General MacArthur. Adams gave his impressions of MacArthur in a contemporary letter:[6] "MacArthur talked freely and gave us his ideas on Japan—quite revealing. He's no doubt a remarkable individual, sincere, idealistic but practical and [it] is easy to see the showmanship he's reported to have. I was impressed. He knew how to be cordial and appealed to all of us."

The report of the group[7] gave a succinct account of the formal organization of science in Japan, of the nature of the universities, of the status of research in them and in industry, and of the necessary role of science and technology in rebuilding the shattered Japan economy. The Japanese universities combined the disadvantages of the German university system with few of the advantages. The feudal and authoritarian nature of Japanese society and science was antithetical to Adams's most deeply cherished principles of society and education—and doubtless to those of his colleagues also. The report recommended reorganizing Japanese science along more democratic lines and making the universities more open. It encouraged more interchange among universities, and between the universities and industrial research through consulting connections with professors, through industrial support of fundamental research in universities, and through making industrial research positions more attractive to university graduates. "Science" in Japan included the humanities and social sciences as well as the physical and biological ones. The report recommended the formation of national professional societies in fields where they did not exist and emphasized the need for adequate libraries, laboratory equipment, and funds for research.

Thus the report and Adams's speeches, parts of which were included directly in the report were expressions of Adams's credo. The following quotation[8] recalls Adams's German experiences with the one-professor system, and the completely different system he created at Illinois:

> Conceivably, the prevalence of the "chair" system [the German professoriate] within the public universities may be an important factor in the imperfection of communications both between and within universities. The idea that a full-fledged university properly consists of a given number of facilities, that each faculty is subdivided into chairs, and that each chair is a kingdom in which the professor exercises jurisdiction over an assistant professor and one or more assistants, is assuredly alien to American thought.

After reciting the pernicious effects of this system, the report discusses graduate training:[8] "In general, graduate study in Japan seems to involve attachment to a single professor and work under his direction almost entirely, rather than exposure in advanced seminars to the competition with other students or to the instruction of other professors."

The report as a whole is an impressive document, representing the distillation of many years of experience of Adams and his colleagues. They carefully avoided the appearance of handing down scientific commandments to the Japanese, although their recommendations were made frankly, even bluntly, and they freely acknowledged that differences in cultural patterns might make American models only partially successful in Japan. Adams dealt again with the concept of democracy in science in his second speech, given at the residence of the prime minister at the inauguration of the Renewal Committee on August 25, 1947, and also, in ceremonial style, warmly thanked the Japanese for their hospitality and cooperation during the visit.

The report was received by the Japanese with indications of gratitude to the group for the substance and the style of their visit. A letter of transmittal signed by MacArthur endorsed the report and expressed appreciation[9] "to the members of the Scientific Advisory Group for the time they so generously gave and the contributions they have made for improving research in Japan. They have earned the thanks of the Allied Nations for their extremely valuable report." President Richards of NAS (successor to Jewett) wrote Adams,[10] "It is perfectly apparent to me that you did a swell job; that a better group couldn't possibly have been selected and that both the Academy and the Military are to be congratulated on your performance.... I am particularly glad that the interest of the Mission was sufficient to outweigh the discomfort of the heat and humidity."

Of all of Adams's public service, the Japanese mission was one of the most important and probably afforded him great personal satisfaction. It is clear from his later visits to Japan and from the comments of Japanese scientists (even when the Japanese rhetoric is discounted somewhat to make it comparable to American modes of expression) that Adams was regarded as a father figure by Japanese scientists, particularly the younger ones whose cause the report championed. Adams was venerated by Japanese chemists for the rest of his life.

Beyond the personal popularity of Adams and the respect that the bearing and attitude of his group inspired in Japan, it is important to inquire how fundamental and permanent the results of the report recommendations were. They certainly opened to the Japanese a totally new view of research, teaching, and the role of fundamental science in supporting technology. Positive testimony is furnished by K. Nakanishi of Columbia, who wrote:[11]

> He [Adams] was one of the earliest Americans to visit Japan after the war. He was a member of an official delegation and it is Roger Adams who was responsible for completely overhauling the Japanese education system and changing it to the current format which closely follows that of the U.S. This means the 6–3–3–4 years of education plus the additional 2 years for the masters degree and an additional 3 years for Ph.D....
>
> To answer your question in summarizing he certainly had a great indirect influence in me being at Columbia today. However, more importantly if it were not for him probably the Japanese education and research system would have been different than it is now.

Nakanishi's estimate was confirmed by T. Takeshita of Du Pont,[12] who said that the present university system is modeled after the American one; the universities have been democratized (perhaps too much) and the university system has been broadened and is less elitist than it was. Other qualified observers do not see as marked a change in the Japanese university system as described above, but a balanced view of the available evidence does indicate a beneficial effect of the Adams mission on Japanese higher education.

The success of the first mission led General MacArthur's office to request NAS to send a second scientific mission to Japan during the election of Japan's first

National Science Council. The second committee was chaired by Detlev Bronk, foreign secretary and soon to be president of NAS; the other members were Adams, the only representative of the first mission; E. C. Stakman, biochemist of Minnesota; Zay Jeffries, engineer and vice-president of General Electric; and I. I. Rabi, physicist of Columbia. The tentative schedule called for them to be in Japan from November 26 to December 17, 1948.[13]

The army furnished transportation, and its first travel directive to Adams was so involved and inconvenient, requiring him to leave from Springfield, Illinois, a 100-mile bus ride from Urbana, that it evoked a coldly furious letter from him. In this letter, which was never sent, he expressed his accumulated frustration with military red tape both in general and in detail. As he pointed out, a group of leading scientists was being asked to donate their time at an inconvenient season of the year and should be treated as well as a group of brigadier generals. In a few days, however, reasonable transportation from Chicago to the West Coast was arranged. The adventures in travel were not over yet, as one of the four engines of their flying boat failed 600 miles from San Francisco and the plane was obliged to return under air escort to the mainland.[14] The rest of the trip went according to schedule.

Adams was naturally anxious to see how effective the first mission had been in influencing Japanese science and was evidently well pleased with what he found. In a letter to President Richards of NAS of January 3, 1949, he said:[13]

> I am sending you just a line to tell you my impressions of the Academy's scientific mission to Japan. Perhaps I am in a better position to judge than the others who made the trip for the first time. I recall that in 1947 I was considerably in doubt about the accomplishments of the group, but before a year passed was convinced that we had fulfilled our obligations satisfactorily and that the results were effective. I feel the same way about this shorter trip just completed. A memorandum was submitted to General MacArthur which General Marquat and Dr. Kelly were very anxious to see done. The report for the Japanese is not yet entirely ready.
>
> Bronk was an excellent chairman, as we were all sure he would be.... Bronk is certainly a man with plenty of physical stamina and in spite of a full program, a bad cold and very little sleep he held up throughout without a complaint. The five of us were very congenial at all times and I myself enjoyed the trip immensely.

Years later Adams, in a letter to a correspondent who was preparing a biography of Zay Jeffries, gave a description of the group's interview with MacArthur:[15]

> ...We had luncheon and a long session with General Douglas and Mrs. MacArthur at the Embassy. There were one or two other military representatives, but other than that it was a private luncheon with the discussion lasting until well after four in the afternoon. The occasion was a most delightful one.

> You are quite right that the afternoon consisted primarily of a monologue by General MacArthur. As I recall, we didn't have more than two or three questions that we could slip in from time to time; and that was adequate to keep the monologue going. I don't recall what questions Zay may have asked the General, but I am sure one of the questions came from him.
>
> The committee, in general, was inclined to be prejudiced against the General before we had met him and talked with him; but after our long conversation we were convinced that he was exceptionally well informed and had sound ideas both with respect to managing the Japanese but also taking care of United States interest. What impressed Zay most, and he spoke about it aferwards, was the comprehension that the General had of American business and how an organization had to be run in order to be successful.

In an informal report on the mission to NAS at its business meeting on April 26, 1949, Bronk reported that the group traveled widely in Japan and had a long interview with MacArthur and also with the emperor. The only record of the meeting with Emperor Hirohito appears to be a newspaper interview with Adams after his return:[16]

> The visit included a 40-minute interview with the emperor, which consisted primarily of "exchanging niceties. But it was interesting to see him."
>
> The party saw the emperor at his palace, with two members of his household. The group, after exchanging the "niceties," talked about biology, then saw the emperor's collection and "nice" laboratories. The group was also entertained by the prime minister at tea.

Bronk reported further that the mission had long discussions with the Japanese Renewal Commission, with representatives of scientific organizations, and with various agencies of SCAP. Members of the mission gave thirteen scientific lectures to specialized scientific societies and had "many informal conferences and conversations with scientists as individuals."

Bronk felt that the mission encouraged Japanese scientists, helped to make them aware of the status and recent developments of science in the United States, and increased the general prestige of the new Science Council of Japan. He also beleived that Japanese science would be oriented toward America instead of Germany, as it was before World War II, and he emphasized the responsibility of SCAP and American science in encouraging the new direction of Japanese science.

General Marquat of SCAP wrote to Jewett on December 31, 1948, of the second mission: "I am again reminded of our debt to you for your early appreciation of the need for the highest equality men to advise us on the problems which exist here....The second group, as the first, performed their duties in a manner that gained the highest praise and respect of all who came in contact with them. The recommendations to be given by the second group will be studied most carefully and every effort will be made for their practical implementation." What these recommendations were in detail is not evident from documents at hand. Adams later felt that SCAP imposed the whole American educational system too rigidly on Japan. As a practical if not official sociologist, he was well aware that cultural institutions can be exported only to a limited extent:[17]

> Americans are proud of their form of Government, their democratic but inefficient system of education, their skills in science, their technological know-how, they naturally stress the adoption of the systems they are acquainted with and understand, even though these systems are not the most adaptable to another country with different traditions and customs. I looked with dismay, when I visited Japan, upon the policy of the educational division of the United States Military Government of persuading the Japanese in a friendly but firm way to change completely their college and university system from the European which has always been in force to the American type. It was a most difficult task for the Japanese for they had neither the space, the personnel, nor experience. A state of confusion resulted which will take years to overcome. I personally believed the European system much more suitable for the Japanese people than the American system and that by relatively minor modifications of the former, a more democratic educational program could have been developed with greater effectiveness and with less delay and expense. The educational representatives of the United States, however, were acquainted thoroughly only with the American system and to them it was the ideal. Recommended changes in diet, clothing, living conditions, sanitary and medical services cannot be effectively instituted in a backward country in a sudden revolutionary way. They must be evolutionary from within without sudden change in the customs which a country has already adopted.

Adams's continuing interest in Japan was shown by several subsequent trips as well as by extensive correspondence with Japanese chemists. The outstanding research published by Japanese organic chemists after 1950, of much higher quality than that before 1940, undoubtedly owes something to his informal activities and interest in Japanese chemistry as well as to his official missions.

Literature Cited

1. R. Connor to DST, January 24, 1979.
2. Patterson to Jewett, June 17, 1947, in NAS Archives: NAS: Sc Adv. Group on Sc in Japan; First; abbreviated hereafter as NAS Japan I. Preliminary conversations between Patterson and Jewett took place a couple of months earlier, loc. cit., Jewett to Patterson, June 19, 1947.
3. Relevant correspondence in Illinois Archives, President Stoddard File 2/5/15, 364. Jewett's formal letter of invitation to RA is printed in the report of the committee, pp. iv–vi, NAS Japan I.
4. "Preliminary Draft Itinerary for Group," NAS Japan I.

5. RA to Jewett, September 30, 1947, NAS Japan I. A handwritten letter, RA to A. N. Richards (the new president of NAS) of September 6, 1947, from Adams's summer home in Greensboro, Vt., says he is "trying to cool off and make up sleep before a strenuous ACS meeting in NY next Friday"; ibid.
6. RAA, 7, Mrs. RA & Lucile, Japan 1947; RA to Mrs. Adams, July 22, 1947.
7. *Reorganization of Science and Technology in Japan*, Report of the Scientific Advisory Group to the National Academy of Sciences, Tokyo, Japan, Aug. 28, 1947; NAS Japan I.
8. Report, p. 37.
9. Statement of March 4, 1948, signed by Douglas MacArthur; NAS Japan I.
10. A. N. Richards to RA, September 10, 1947; NAS Japan I. A letter from General Marquat of MacArthur's staff was equally enthusiastic, praising "their conscientious attention to duty under arduous climatic conditions and also the wide human understanding they displayed"; W. F. Marquat to A. N. Richards, September 4, 1947, NAS Japan I.
11. K. Nakanishi to DST, May 22, 1968.
12. We are indebted to R. M. Joyce for obtaining the views of Takeshita.
13. All documents relating to this mission, unless otherwise indicated, are from NAS Archives, NAS: Science Advisory Group on Science in Japan: Second.
14. Clipping from the *New York Times*, November 23, 1948, from archive above.
15. RAA, 37, M; RA to W. D. Mogerman, May 23, 1968.
16. RAA, 57, Press Clippings, 1940–1949; interview with Adams in a Champaign–Urbana newspaper, December 30, 1948.
17. RAA, 54, Speeches AAAS; speech to AAAS, December 27, 1950.

Illinois and Research, 1943-67

Researches, 1943–61

Adams's research students continued work on his problems during part of the 1940–45 period, when he was spending most of his time in Washington. They gradually shifted, however, to war research projects, such as Marvel's program on synthetic rubber or the synthesis of possible antimalarial drugs.

The decrease in Adams's research program in these years is shown in Table I. The surprising thing about these figures is not the decrease, which was inevitable because of Adams's Washington work, but the speed with which he revived his program after his return to Urbana in March 1946. In 1949 he was sixty years old; his Japanese trips had used up several months, and he was still carrying a heavy burden of editing, public service activities, and consulting. His interest in research and his students was undiminished; from 1943 to 1957, inclusive, he trained fifty-eight Ph.D.'s and wrote over 100 scientific papers, in addition to numerous lectures and general publications, such as the one with Harold R. Snyder on "Academic Opportunities" and his own paper, "Responsibilities of Employers to Professional Employees."[1] His research in these years centered on the structure of natural products, particularly the *Senecio* alkaloids, but also on other heterocyclic natural products. He continued his work on hindered rotation and also studied intensively a little-known class of compounds, the quinone imides, for which he discovered a general synthesis. The postdoctoral fellows, attracted by his reputation, played an important role in reestablishing his research program.

The instrumental revolution, including the applications of several branches of spectroscopy, changed fundamentally the methods of establishing chemical structures. Adams had made use of ultraviolet absorption in his earlier work; after 1945, when instrumentation became generally available, he used infrared spectra (IR) routinely, and in his last papers he utilized nuclear magnetic resonance (NMR). His papers indicate that he personally was not overly familiar with these powerful aids and probably depended on his postdoctoral fellows for interpretation of spectra. Later papers show use of chromatographic separations, including paper chromatography, which were particularly useful in dealing with the complex mixtures of the *Senecio* alkaloids and their degradation products.

It is characteristic of Adams's detached view of himself, rare in people of his eminence, that he realized clearly the cost to his scientific work of his long absence from the university. He wrote Gilman from Germany in 1946,[1] "I suppose by this time you are hard at work on academic problems. I hope to be [also] myself within the next couple of months but whether I shall be able to function again as a research director is questionable. I have done about everything else for the past six years and I shall probably not realize how much I have forgotten until I dig in once more."

Physical organic chemistry, and especially the study of reaction mechanisms, had developed rapidly in this country in the 1930s, particularly in the laboratories of J. B. Conant, L. P. Hammett, P. D. Bartlett, H. J. Lucas, C. R. Hauser, M. S. Kharasch, and their students. By 1950 the younger chemists doing structural and synthetic work thought about reactions and rationalized their results very largely in terms of reaction mechanisms. The theories of conjugated systems, particularly aromatic ones, were based on the ideas of π electron delocalization ("resonance," in one formulation). Charles C. Price introduced physical organic chemistry to the Illinois department in 1937; Elliot R. Alexander continued the process, writing a short book on one part of the field,[2] and, although his career was cut short by a tragic air crash, his successor David Y. Curtin carried on the work on physical organic chemistry.

Although Adams utilized the new instrumentation in his work, he never became at home with the ideas of reaction mechanisms, and his papers after World War II therefore had a somewhat dated quality. The experimental work was still done with his familiar polish, the papers were written with his usual clarity, but his work in general was no longer in the forefront of progress in the field. The surprising thing is that Adams maintained such a large and active research program until he retired and that his work, particularly on natural products, continued to be valuable.

His continuation of the earlier work on the alkaloids from many plants of the *Senecio* and *Crotalaria* groups (over thirty papers) clarified the structures of most of the compounds on which several other laboratories were working. Hydrolysis of the alkaloid gave a "necic acid," which constituted the structural problem, and Adams determined structures for several "necic acids." In one of his last papers,[3] Adams and Nair, by using NMR, made a minor correction for an earlier structure they had postulated for riddellic acid, obtained along with retronecine by hydrolysis of riddelliine, an alkaloid described earlier from *Senecio ridellii*.[4] Other minor corrections in Adams's structures for the various necic acids were made by his laboratory or others. He also established the absolute configuration of asymmetric centers in the retronecine moiety; this connected his long-standing interest

Illinois University Archives

Studying molecular models (1954).

Table I. Adams's Ph.D's and Publications, 1943–50

Year	Ph.D.'s	Publications
1943	6	6
1944	7	4
1945	1	4
1946	0	0
1947	3	3
1948	2	6
1949	7	16
1950	5	17

in stereochemistry with modern correlation of absolute configurations.[5] This paper suggested that the biogenesis of the necic acids involved acetate and acetyl coenzyme A units,[6] whose importance in the biogenesis of polyterpenes, steroids, and other natural products had just been demonstrated by Bloch, Lynen, Cornforth, and others.

Adams's continued research on the active principles in marijuana prepared and tested synthetic compounds of high activity derived from tetrahydrocannabinol.[7] Unfortunately, his marijuana work ended before new techniques of structure determination and separation allowed a complete structural definition of the active compounds in marijuana.

During this last period Adams pursued another theme of his earlier research: the occurrence of optical activity due to hindered rotation around a carbon–nitrogen bond instead of a carbon–carbon bond as in the biphenyls. Some thirty papers appeared in this series. An example below has two centers of hindered rotation and exists in *cis* and *trans* forms; the latter is resolvable into optically active forms, but the *cis* has a plane of symmetry.[8]

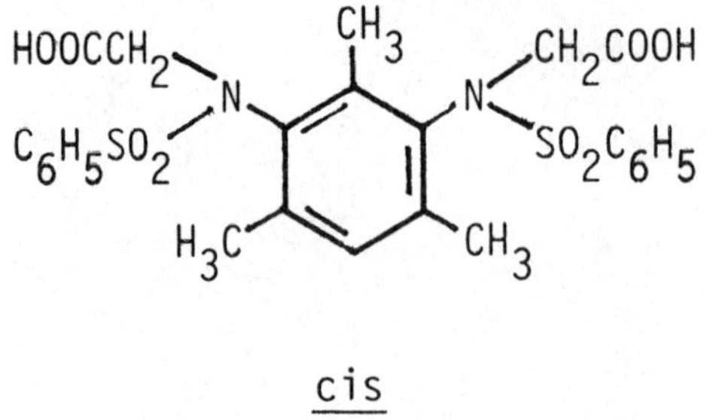

cis

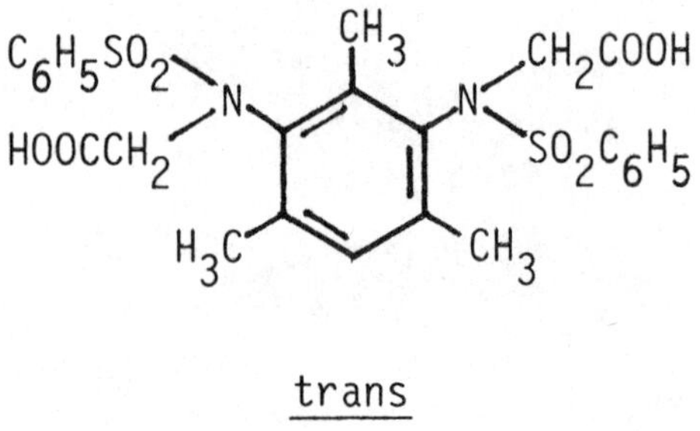

trans

Some measurements of half-lives for racemization were made in compounds of this type, and the results in general followed those in the biphenyl series, although they were not as clear-cut. Activation energies and other derivable quantities were not obtained.

Adams studied in great detail the novel quinone imides, usually the disulfonyl or diacyl derivatives. Adams and Nagarkatti[9] found that oxidation of 1,4-disulfonamidobenzene with lead tetraacetate or red lead (Pb_3O_4) in acetic acid gave the disulfonimide in high yield as a stable yellow compound. The oxidation reaction was a general one yielding disulfonimides or diacylimides from a variety of benzoquinones, naphthoquinones, diphenoquinones, and others.

$NHSO_2R$ / $NHSO_2R$ $\underset{HBr}{\overset{Pb(OAc)_4}{\rightleftharpoons}}$ NSO_2R / NSO_2R $\xrightarrow{HCl}$ $NHSO_2R$ / Cl / $NHSO_2R$

The compounds behaved like quinones, were reduced to the starting material by weak reducing agents like HBr, and added a wide variety of reagents to the quinonelike conjugated system, regenerating the aromatic ring, as Adams showed in many later papers. Among the adducts were HCl, HN_3, RSH, sodiomalonic ester, amines, benzenesulfinic acid, and conjugated dienes in a Diels–Alder reaction. Monosulfonimides were synthesized.[10] Some of the addition products obtained from quinone imides could be converted to heterocyclic systems, such as indoles, benzofurans, and others. Although these new heterocyclic syntheses appeared useful only in special cases, he considered them excellent teaching problems for research students, because the reactions usually worked and frequently gave mixtures whose separation was good experience. As he did with biphenyls, he continued this series to the point of diminishing returns in scientific interest.

$NHSO_2R$ / OH $\xrightarrow{[O]}$ NSO_2R / O

The alkaloid leucenol, isolated from the seeds of a tropical tree similar to acacia,[11] was shown by Adams to be a hydroxypyridone by degradation and synthesis.[12] The leucenol work led him to studies on reactions of pyridine derivatives and pyridine-*N*-oxides; this resulted in interesting rearrangements and synthetic reactions.[13]

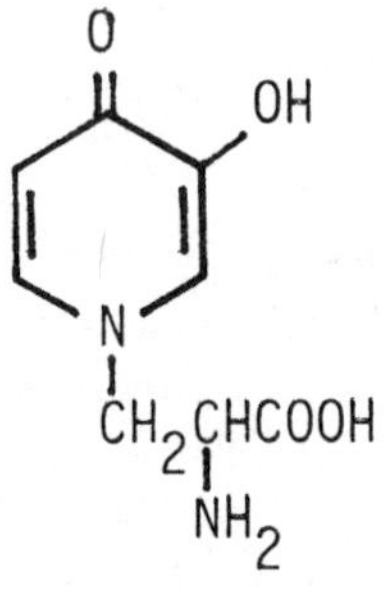

leucenol

In addition to these main problems, the Adams group worked on several other projects, which remained unfinished but laid the groundwork for other research workers to complete with the new instrumental techniques. There was an ingenious but unsuccessful attempt to demonstrate restricted rotation around an aliphatic carbon–carbon bond.[14] The structure of the natural product helenalin, which was found to be an azulene lactone, was examined,[15] and years later one of Adams's postdoctoral students, Werner Herz at Florida State, with abundant help from high-resolution NMR, completed the structural work.[16] Adams investigated briefly some of the "amaroids," quassin and related compounds ["bitter principles" isolated from quassia wood years earlier by E. P. Clark of the U.S. Department of Agriculture[17]], and workers in England, Canada, and Australia eventually solved the structures, again with NMR playing a key role.[18]

Adams never published all the work of his last postdoctorates, who were supported in his retirement by the Sloan Foundation grants; he explains the reason in a letter of 1965 to an overseas friend:[19]

> There are many compensations in being retired. In the United States one is no longer permitted to direct student theses; so that it is necessary to hire assistants if one desires help with his researches. I was given money from one of the foundations to employ postdoctorates and I did so for about four years. I then became so involved with various other activities that I was neglecting my reading and consequently felt that I was not being fair to my associates. Moreover, one has to be an expert in interpreting IR, NMR, and mass spectrographs [sic] and the like; and it was perfectly obvious that I was going to have great difficulty in competing with the younger staff men.

Adams's scientific publications spanned the years 1910–61, although his second paper did not appear until 1916. As his active research and teaching career drew to its close, the more reflective observers of the chemical scene realized that an era in American organic chemistry was ending. Few people had contributed so much for so long to so many different facets of the field. American organic chemistry had solid foundations when Adams started his work; during his lifetime it came of age and gained international stature. His own contribution was a major factor in this rise, and at his retirement American organic chemistry was on a level of accomplishment as high as or higher than that of any other country.

LITERATURE CITED

1. H. R. Snyder and RA, *Chem. Eng. News, 36,* January 27, p. 66 (1958); RA, ibid., *23,* 1706 (1945); RA to Henry Gilman, January 21, 1946.
2. E. R. Alexander, *Principles of Ionic Organic Reactions,* Wiley, New York, 1950.
3. M. D. Nair and RA, *J. Am. Chem. Soc., 83,* 922 (1961).
4. RA and B. L. VanDuuren, ibid., *75,* 4638 (1953); RA, K. E. Hamlin, C. F. Jelinek, and R. F. Phillips, ibid., *64,* 2760 (1942).
5. RA and D. Flès, ibid., *81,* 4946, 5803 (1959).
6. RA and M. Gianturco, ibid., *78,* 4458 (1956).
7. RA, M. Harfenist, Jr., and S. Loewe, ibid., *71,* 1624 (1949).
8. RA and J. J. Tjepkema, ibid., *70,* 4204 (1948).
9. RA and A. S. Nagarkatti, ibid., *72,* 4601 (1950); this paper gives references to Willstätter's earlier work on quinone imides. Adams published 46 numbered papers in this series, of which the above is I, and in addition six papers on closely related compounds.
10. RA and J. H. Looker, ibid., *73,* 1145 (1951).
11. RA, S. J. Cristol, A. A. Anderson, and A. A. Albert, ibid., *67,* 89 (1945).
12. RA and J. L. Johnson, ibid., *71,* 705 (1949).
13. RA and I. J. Pachter, ibid., *76,* 1845 (1954) and later papers.
14. RA, R. S. Voris, and L. N. Whitehill, ibid., *74,* 5588 (1952).
15. RA and W. Herz, ibid., *71,* 2546, 2551, 2554 (1949).
16. W. Herz et al., ibid., *85,* 19 (1963); W. Herz and H. B. Kagan, *J. Org. Chem., 32,* 216 (1967).
17. RA and W. M. Whaley, *J. Am. Chem. Soc., 72,* 375 (1950).
18. Z. Valenta et al., *Tet. Letters, 20,* 25 (1960); R. M. Carman and A. D. Ward, ibid., *1961,* 317; *Austral. J. Chem., 15,* 807 (1962).
19. RAA, 34, European Correspondence, S.; RA to C Schöpf, Jan. 29, 1965.

The Illinois Department, 1946–67

A general view of the Illinois department from 1946 to 1967, when it celebrated its centennial, rounds out the story of Adams's last years as an active faculty member and as emeritus professor.[1] The department continued to grow in numbers of graduate students, Ph.D.'s, faculty, and undergraduates. Table II shows the doctorates awarded (the numbers of earlier years are given in Table II, page 95). The annual number of Ph.D.'s continued fairly constant after 1949 at about sixty. The percentage of organic Ph.D.'s relative to the total granted by the Illinois department remained about constant over ten-year intervals from 1930 to 1960, but there were large annual fluctuations. The percentages of organic Ph.D's, by decades, were forty-seven percent (1930–40), fifty-six percent (1940–50), and forty-seven percent (1950–60); for 1961–67, inclusive, the figure dropped to thirty-three percent.

Adams's years as head of the department from 1946 to 1954, when he became research professor, were in some respects anticlimactic. During his long absences in 1941–45 and his subsequent missions to Germany and Japan, American society had altered markedly, and the academic scene was no exception. The postwar complications of large increases in numbers of undergraduates and graduate students and other social changes increased the burdens of university administrators. Another phenomenon was the large proportion of married graduate students after 1945. Although Adams, like most people familiar with the graduate scene before 1940, did not like this situation, it had to be accepted.

Outside research funds after 1945 from both government and private agencies made more graduate students and postdoctorates available, increasing the research activity quantitatively and qualitatively. However, these funds brought serious administrative and policy questions. Faculty members had to spend much time in writing proposals and reports to obtain research grants and contracts and in reviewing proposals from other investigators; the flexibility and continuity of research programs were frequently adversely affected by the uncertainty of continuous funding; and the accounting of funds required increased administrative staff. In spite of these and other disadvantages, outside research funds were indispensable for a department to maintain a leading position.

The instrumental revolution required large sums for major pieces of equipment, such as mass spectrometers, NMR spectrometers, X-ray equipment, and computers, which were generally available only from government grants. Although these instruments revolutionized research by increasing the speed of getting results and by opening up entirely new areas to investigation, they posed serious problems to Adams and his successors and to their colleagues in all major universities. Aside from the large initial costs, the new instrumentation required highly skilled workers to use and maintain the equipment efficiently. The days were past when good research could be done with simple equipment like Claisen flasks and melting point baths.

Illinois University Archives

Leaders for over fifty years of Illinois chemistry: H. E. Carter, Adams, H. S. Gutowsky.

Evidence, both written and oral, suggests that before 1954 the younger staff members were becoming restive during the last years of Adams's headship. This occurred in spite of the fact that Adams continued to give his junior colleagues every chance to develop their own research careers. In the 1930s, decisions were made by Adams in consultation with the division heads; this informal, off-the-cuff manner of operation was no longer adequate after 1946, both because of the complexity of the problems and because of the wish of younger staff members to be heard. The 1930s represented a remarkable peak of achievement for Adams and

Table II. Ph.D's from Department of Chemistry and Chemical Engineering, 1946–67

Years, Inclusive	Ph.D's[a]
1940–4	240
1945–9	218
1950–4	303
1955–9	260
1960–4	305
1965–7	192

[a]See Ref. 2.

his contemporaries, but the department changed inevitably as younger men came to the fore.

Adams's contemporaries on the faculty were approaching retirement also. Marvel, whose work in polymer chemistry had given Illinois a unique position in this important field, retired from Illinois in 1961 but continued a very productive research program at Arizona for many years. R. C. Fuson retired in 1963 and moved to the University of Nevada; W. H. Rodebush and W. C. Rose stepped down in 1955; and there were other key retirements in analytical and inorganic chemistry.[1]

In organic chemistry C. C. Price moved to Notre Dame in 1946, and his place was taken by E. R. Alexander, Cope's student.[3] Alexander showed great promise in physical organic chemistry until his untimely death, and he was succeeded in 1951 by D. Y. Curtin, a former student of Price. Curtin had been a postdoctorate with P. D. Bartlett and came to Illinois from the faculty at Columbia. N. J. Leonard, Rhodes scholar, Ph.D. with R. C. Elderfield at Columbia, and a postdoctorate with Adams, was reappointed to the faculty in 1946. H. R. Snyder, a student of J. R. Johnson and Adams's postdoctorate, prosecuted studies in synthetic and structural organic chemistry. Later Leonard worked in this field and developed very fruitful interests in biochemistry. Younger able colleagues in organic and other divisions were added constantly.

Adams paid eloquent tribute to his chief colleagues of the golden years of the 1930s. He wrote in 1966 to Ralph Shriner:[4]

> The years from 1927 to 1941, while you were with us, and Speed, Bob [nickname of R. C. Fuson], you and I were the principal organic staff members, were to me the most satisfying years during my service as head of the department. We all worked so well together; and in spite of being plagued for a long period by the depression and shortage of funds for operation, the morale of staff and students was high. Each one was striving to accomplish as much as he could, and the record of achievement was noteworthy.

This rather wistful tone, which was as close as Adams ever came to nostalgia, sounds again in this tribute to Marvel:[5]

> He was a major contributor to the success of the Department during those many years [1920–61]; in teaching, in research and in administration. But of even greater significance to the Department was his interest in the welfare of the students. His genial personality attracted them and he was kept busy counseling not merely his own students but very many others. His popularity circulated nationwide and served to attract to Illinois many able new students who were about to begin graduate study.... Speed's achievements in research have resulted in his enviable international reputation. Whenever a symposium is held involving the chemistry of polymers he is the first man to be invited to speak.
>
> Very few chemists have as many friends as Speed or have enjoyed so much success in so many phases of academic life.

Illinois University Archives

Adams and Carl S. Marvel (1960).

Illinois University Archives

The first section of the Roger Adams Laboratory.

The Noyes Laboratory had been overcrowded and unsafe for many years; Adams's reports to the Illinois luncheons at ACS meetings routinely recounted lack of progress in renovating or extending the building. For years a new building for chemistry had been planned and was vigorously advocated by Rose in 1943, when he was acting head. With Adams's support,[6] Rose forcefully presented the deficiencies in safety, ventilation, and space. None of these needed emphasis to the many hundreds of research students and thousands of undergraduates who studied in the Noyes Laboratory.

In 1951 the new East Chemistry was completed, but rising building costs allowed space for biochemistry and chemical engineering only. In 1967 an addition to East Chemistry was finished, which housed organic and analytical chemistry. It was a far cry from the 1902 section of the Noyes Laboratory with its wooden frame and unworkable hoods to the admirable laboratories in East Chemistry.

In recognition of the real situation, the chemistry department was renamed in 1953 as the department of chemistry and chemical engineering, the notice to the staff announcing the change being signed "Roger Adams, Head—Chemistry Department."[7]

The logical successor to Adams as head was Herbert E. Carter (Ph.D. 1934, Illinois) distinguished for his research in biochemistry and for his administrative talent. Some of his problems as the new head arose from the increase in graduate student enrollment over the period 1954–66 from 259 to 430, and from an increase in the undergraduates taught per semester from 4043 to 4936.[1] Adams's magnanimity prevented a possible problem: after he stepped down, he did not attempt to dictate policy to the department or to Carter. As Carter wrote,[8] Adams never visited him in the departmental office to talk about department policy, but when Carter went to Adams's office, the latter felt free to discuss anything about the department in detail. Adams adhered fairly strictly to this self-imposed protocol and honored the right of his successors to develop the department in their own way. Nevertheless, Adams was too positive a personality to keep his views to himself completely and he volunteered his ideas on department policy occasionally in letters or memoranda. He considered that senior established investigators should have been recruited to replace him and his contemporaries, and he wrote in 1964, "The Chemistry Department at the University of Illinois rates lower among the chemistry departments of the country than it did ten or twelve years ago."[9] This was a reversal of his earlier position of 1944, when he maintained that it was better to develop outstanding people within the university rather than to bring in established workers.[10]

In structural and synthetic organic chemistry, the Illinois primacy was no longer uncontested. In one sense, Adams and his colleagues had been almost too successful as teachers and trailblazers, because their competition was beginning to catch up to them. E. J. Corey, whom Adams recruited in a special trip to MIT,[11] published outstanding work at Illinois from 1951 to 1959, when he moved to

Harvard. In physical organic chemistry the Illinois work was excellent but was not as widely recognized as that in several other universities.

After 1946 the divisions of inorganic, physical, and analytical chemistry challenged the predominant role of organic chemistry at Illinois with their increased excellence and quantity of work, and biochemistry under H. E. Carter continued the high standard set by W. C. Rose. This reflected a general trend throughout the country as a result of the tremendous impetus in these fields from World War II research, especially from the Manhattan (atomic bomb) Project.

In 1957 C. K. Ingold, the outstanding British physical organic chemist, wrote Adams after a visit to Urbana, thanking him and Mrs. Adams for the party at their "charming home." He continued that he was struck by the advance of the University of Illinois, the "bustling vitality" of the chemistry department, the "quick wittedness" of the students, and the spirit of the department, "of which I have not often seen the like." Ingold's comments are particularly noteworthy because the Illinois and Adams tradition of structural and synthetic organic chemistry was far from the field of Ingold's own interests in physical organic chemistry.[12]

Carter led the department successfully until 1967, when he became vice-chancellor of the Urbana campus. His successor as head was the physical chemist Herbert S. Gutowsky (Ph.D. 1949, Harvard, with Kistiakowsky), with a distinguished research career, particularly in nuclear magnetic resonance. In 1970 the department was renamed the School of Chemical Sciences, with Gutowsky as director. Adams had urged President Willard in 1936 to make this change in status of the department, but it was unacceptable to Willard.[13] The new name described more accurately the very large and varied department of 1970.

In spite of the striking changes in physical accommodation, types of research, and faculty personnel, the Illinois School of Chemical Sciences continued to maintain its tradition of hard work, outstanding performance, and high morale established by William A. Noyes and Roger Adams.

In June 1972 the university named the East Chemistry complex the Roger Adams Laboratory in a ceremony attended by his daughter and her family. Young Roger, Adams's grandson, unveiled the bronze plate that marked the laboratory. Now the buildings named for the two great founders of the Illinois School of Chemical Sciences, William A. Noyes and Roger Adams, devoted to the tradition of outstanding teaching and high scholarship, face each other across Matthews Street.

Literature Cited

1. Much information is given in *Centennial 1967, Department of Chemistry and Chemical Engineering,* University of Illinois, Urbana.
2. The figures for 1940–62 are from *Doctorate Production in United States Universities 1920–1962,* Publication 1142, National Academy of Sciences–National Research Coun-

cil, Washington, D.C., 1963, p. 98. These are limited to chemistry and are slightly smaller than those in Ref. 1, which include doctorates in biochemistry and chemical engineering. The figures for 1963–67 are taken from the lists in Ref. 1, although these figures need a small correction for the reason described above. The figures on the percentage of organic Ph.D's are through the kindness of Elsie Wilson and H. S. Gutowsky.
3. Alexander's name is unaccountably absent from the faculty list in Ref. 1.
4. RAA, 26, Org. Synth. 1962–66; RA to R. L. Shriner, May 10, 1966.
5. RAA, 37, M; RA to J. F. Miller, March 13, 1970; autobiographical article, C. S. Marvel, *Chemtech.*, *10*, 8 (1980).
6. W. C. Rose to President A. C. Willard, August 12, 1943; Illinois Archives 15/1/1, 33, Chemistry 1943–44. This is a five-page letter with two additional pages of tables of student enrollments and estimates of the value of the equipment in the Noyes Laboratory totaling $500,000.
7. Department notice of March 24, 1953.
8. H. E. Carter to DST, March 28, 1979.
9. RAA, 31, Brichford; memorandum by RA.
10. RA to Dean M. T. McClure, September 4, 1944; Illinois Archives Deans Office, 15/1/1, Chemistry 1944–5.
11. RAA, 36, C; E. J. Corey to RA, March 17, 1971.
12. RAA, 7, Ingold; C. K. Ingold to RA, November 3, 1957.
13. Confidential letter, RA to A. C. Willard, February 28, 1936; Illinois Archives, Willard, 2/9/1, 20, LA & S, Chemistry. Reply, ibid., March 6, 1936. Willard did not think the change would solve any problems and would cost more money.

Broader Horizons

Outside Service, 1946–71

An earlier chapter recounted Adams's manifold activities in addition to his teaching, research, and administrative work in Urbana from 1918–42. From his return to Urbana in 1946 after the years of war service until his death, he shouldered a great variety of responsibilities outside the university. All had in common his aim to improve Illinois and American science and to increase international scientific contacts.

Adams's interest in *Organic Syntheses* remained at a high level for the rest of his life. He continued, even in the war years, to participate in the annual volumes and to make suggestions to the secretary about preparations, meetings, and management of the corporation's growing portfolio of securities.[1,2] An entertaining example of his watchful care followed the regular meeting of the editorial board at the September 1950 ACS meeting in Chicago. Exercized by the cost of the dinner for the group, Adams protested the extravagance to the secretary, Norman Rabjohn (Missouri), urging that future dinners be limited to eight dollars per person including cocktails.[3] After assuring Adams that the dinner was not as expensive as he thought, Rabjohn with some difficulty arranged dinner for seven dollars for the next meeting. When the group sat down to eat, the headwaiter (in on the joke) presented Adams with a special plate containing two hot dogs and some baked beans. Adams viewed his feast with rueful surprise but finally joined in the general hilarity. He made no further complaints about the price of the dinners.[4]

As the years passed younger men were appointed to the editorial board, and a conscious effort was made, starting particularly with John D. Roberts, editor in chief of Volume 41 (1961), to renew a program of solicitation for preparations representing new synthetic procedures and reagents, a precedent followed by succeeding editors.[5] The emphasis on new or neglected general reactions and the inclusion of an account of the significance and usefulness of each preparation, also started by Roberts, maintained the series with unabated vigor.

The value of *Organic Syntheses* was occasionally questioned by some chemists with strong interests in physical organic chemistry. Adams's reply to one such opinion, written when he was seventy-six years old, is significant because it shows his insistence on the value of training in synthetic chemistry:[6]

> My own opinion is that the trends in organic chemistry at the present time actually enhance the value of Organic Syntheses. In many of the larger institutions there is less and less stress upon laboratory practice in organic chemistry as it applies to synthesis. As a consequence, many of the graduate students are more familiar with the new physical tools than with the synthesis of chemicals they may desire. Many relatively rare chemicals are available at the present time, and often the research man is able to acquire what he needs in that way. But when he can't, he must synthesize them. He then finds himself handicapped by lack of experience in synthesis and will often spend much more time than is necessary before he obtains what he needs. Organic Syntheses should be very helpful to such an individual....
>
> ... Organic Synthesis is contributing not merely to the academic but to the industrial chemist. Industrial research directors have informed me more than once that they now have difficulty in hiring men who are adept at synthetic work and that it requires a year or two for them to acquire the training that they used to get in the universities. To be sure, the industry is also very much interested that they have the knowledge that is now taught them in the use of the modern effective tools and in the mechanism of reactions. The pharmaceutical industry and a large part of the chemical industry are still based on synthesis, and certainly Organic Syntheses is helpful to their chemists.

Adams's close attention to the investments of *Organic Syntheses*[7] was accompanied by a rapid increase in assets, which reached nearly $500,000 in the 1960s.[7] He relished arguing with his colleagues about investment policy; his long experience and practical awareness of economic trends made him almost a professional investment counselor. A plan suggested by Roberts and adopted after considerable debate allowed graduate students in chemistry to purchase the annual or collective volumes at half price. Initially there was substantial worry that this plan would greatly reduce the assets of the corporation, but actually the enthusiasm at the graduate level added new impetus to the overall program.

When Adams retired officially at Illinois in 1957, the *Organic Syntheses* officers and editors sent him a letter of appreciation, and acting on N. J. Leonard's suggestion, they established the Roger Adams Medal in organic chemistry with an honorarium of $5,000, later raised to $10,000. *Organic Syntheses* contributed seventy-five percent and *Organic Reactions* the remainder. This award, administered by the Organic Division and presented biennially at the Organic Symposium, had D. R. H. Barton as its first recipient in 1959, when in a moving ceremony the

Receiving the fiftieth volume of Organic Syntheses *from its editor, Ronald Breslow.* Illinois University Archives

original of the medal was presented to Adams. He had taken an active part in the discussions involved in establishing the award, which won out over the idea of a sizable contribution to the American Chemical Society for their new building in Washington; men, not buildings, was Adams's creed. Later recipients of the Adams award, up to Adams's death in 1971, were R. B. Woodward (Harvard), P. D. Bartlett (Harvard), A. C. Cope (MIT), J. D. Roberts (Caltech), V. Prelog (Swiss Federal Institute), and H. C. Brown (Purdue). One of Adams's last letters told Brown that because of ill health he would be unable to attend the Organic Symposium in June 1971 when Brown was to receive the award.

Organic Reactions, projected around 1939 by Adams and some of the *Organic Syntheses* group, was a collection of articles about specific reactions with which the writers had firsthand experience, exhaustive literature surveys, and some detailed experimental procedures. A. H. Blatt and H. R. Snyder served as associate editors, and Adams was president and editor in chief from 1942 (when the organization was incorporated in Illinois) until 1960 when Volume 10 was published. A. C. Cope succeeded Adams until his death in 1966, when W. G. Dauben became president and editor in chief.[8]

As with *Organic Syntheses*, authors of articles in *Organic Reactions* received no royalties, and the editors did their work as a public service. The profits were invested and the assets at one time in the 1960s were over $100,000. Adams devoted much time and energy to reading manuscripts and proof, even when he was in Washington, as well as to the congenial occupation of managing the investments. Not until his last years, when his eyesight began to fail and proofreading became difficult, did his diligence lessen.[9] *Organic Reactions* by its nature involved a smaller group of people and hence was not as sociable an enterprise as *Organic Syntheses*, but both series expressed Adams's interest in organic chemistry, in organic chemists, and in students. Clearly he felt richly repaid for the hard work he devoted to both publications.

One of the most important of all Adams's activities began when Alfred P. Sloan, Jr., planned to extend his contributions to American science. Sloan (1875–1966, B.Sc. 1895, MIT) for many years the president and chairman of the board of General Motors, established a foundation in 1934 for philanthropic work in a variety of areas. In 1954 Alfred Sloan wished to formulate a definite basic research program for his foundation, and on June 8, 1954, Karl Compton, a trustee of the foundation, wrote Sloan a long letter about possible ways in which the foundation could plan its program to aid science. Among other things, Compton acquainted Sloan with Adams's success in research, his ties with industry and with national affairs, his loyalty to Illinois, and his experience on a committee that made a study of the Mellon Institute. He wrote,[10] "I could think of no better person than Roger Adams to take charge of the proposed study. He probably has a wider acquaintance among chemists in this country than any other person."

Sloan took Compton's advice about Adams, but his approach to Adams was singularly diffident and circuitous. It is reported that Sloan actually was shy,[11] and the letters from him to Adams bear this out. He asked General Lucius D. Clay, who had retired from the army and was a member of the Sloan Foundation board, to write an introductory letter to Adams because Clay had known Adams in Germany. In a "Dear Roger" letter, Clay described Sloan's proposal, and Adams, in a "Dear Lucius" letter, agreed to talk with Sloan, though he was hesitant about assuming further activities because of his present commitments. Only then did Sloan address Adams directly, referring to Clay's introduction and his own purpose to support fundamental scientific research. He urged Adams to head a committee to formulate a program. Adams promised to confer with Sloan in New York at the time of the ACS meeting the week of September 4, 1954,[11] but he actually met Sloan there in August, probably on the way to his summer place in Vermont.[12] On August 20 Adams agreed to serve as chairman.[12] Adams undoubtedly visited Sloan at the earliest possible date because of Adams's wish to lose no time in encouraging Sloan to finance basic research in the physical sciences.

Adams selected a committee of four to work with him: M. J. Kelly, president of Bell Laboratories;[13] W. Albert Noyes, Jr., chairman of the chemistry department of the University of Rochester; J. A. Stratton, vice-president of MIT; R. W. King, retired vice-president of the Bell Laboratories.[14] The committee prepared several drafts of a program and met with Sloan November 8, 1954, when he emphasized that if the venture proved successful, he would grant funds beyond the $5,000,000 originally promised.[15]

The committee's plan proposed a full-time program director, a man of broad scientific interests and acumen, and a group of five outstanding research men in the physical sciences to evaluate the research proposals and persons recommended to the foundation. The heart of the committee report represents what Adams had been advocating and practicing all his life:

> It was agreed that in allocating its grants the Sloan Foundation should place emphasis primarily on men—not projects or institutions as such—but that the climate of the educational institution for research should be given consideration. In seeking men, particular heed should be paid to younger men who offer marked promise. It was recommended that Sloan grants should be for use in definitely stipulated areas of research under a specific individual or group of investigators, and not for departmental use broadly.

The remainder of the report emphasized relieving the grantees of as much paperwork as possible and giving them a free hand in their researches. Sloan received the program with enthusiasm, writing Adams,[16] "Let me thank you for your leadership in developing the program now about to take form. I am indeed grateful to you and science should be grateful to you; i.e., if we make good on our program, which I am sure we will."

Adams succeeded in persuading Richard T. Arnold (b. 1913), an Illinois Ph.D. (1937) with Fuson and an outstanding chemist and administrator, to become director of the basic research program in physical sciences thus initiated, a position he held for five years. The program committee to evaluate applications consisted of A. C. Cope (MIT), K. S. Pitzer (Berkeley), F. Seitz (Illinois), A. W. Tucker (Princeton), and J. B. Fisk (Bell Laboratories), all distinguished scientists.

The program developed into an outstanding success, as Sloan had hoped it would. The award of a grant from the Sloan Foundation was a coveted honor for the prestige as well as for the unrestricted research funds it brought. Few programs have done more to develop promising young scientists and to promote outstanding research, considering the relatively modest sums expended. Adams's leadership in planning this program was a major service to American science. In a sociological study on the reward system in American science, it was found that Sloan Fellowships had a "visibility" of 83 and a "prestige score" of 3.18 on a scale in which the Nobel Prize had values of 100 and 4.98, respectively, and membership in the National Academy of Sciences had values of 95 and 4.22, respectively. The Sloan Fellowships scored approximately the same values as an honorary degree from Harvard.[16]

Adams received a grant of $52,000 from the Sloan Foundation to support his own research for the period September 1, 1957, to May 31, 1961, after he retired as research professor emeritus in 1957; he used this money to support postdoctoral fellows and it was renewed with supplements up to August 31, 1967.[17] Postdoctorates employed included A. A. Stingl, S. Miyano, and A. Feretti.[17]

Adams also accepted Sloan's invitation to serve on the board of scientific advisors of the Sloan–Kettering Institute for Cancer Research, and he kept some connection with the institute until his death in 1971.[18] In 1955 Cornelius P. Rhoads (1898–1959), who had been director of the medical division of the Chemical Warfare Service in 1943–45 and knew Adams well, was director of the Sloan–Kettering Institute. Adams served on the Committee on Scientific Policy of the trustees of the Institute from 1954–66.[18,19]

The establishment of the National Science Foundation (NSF) by Congress in May 1950 was the culmination of a long process of proposals and compromise, initiated by Bush's report to President Roosevelt,[20] *Science: The Endless Frontier*, and involving the wartime scientific leaders, Congress, and President Truman. Adams played some part by testifying before a congressional committee, but he was not a leader in the movement.[21] He was a member of the National Science Board of NSF from 1954 to 1960 and conscientiously attended meetings, though he asked not to be reappointed in 1960 because of his age.[22]

Most of the correspondence preserved between Adams and the officers of NSF dealt with the question of taxing the nonprofit research institutes, such as Mellon, Midwest (in Kansas City), and Battelle, of which Adams was a director. In lively exchanges by letters and during board meetings, Adams tenaciously defended his position that such institutes were actually nonprofit and should not be taxed. His point of view was eventually accepted.

The other point emphasized in his correspondence with NSF concerned the establishment of postdoctoral fellowships for young scientists to study abroad. Some of Adams's friends who were well-known professors in European universities informed him that American Fulbright scholars did more traveling than working in the laboratory. Consequently the professors were reluctant to tie up laboratory space for Americans who were taking too literally the Fulbright injunction to travel and get acquainted with the country and people. This low opinion held by European professors of American Ph.D.'s bothered Adams seriously, and he urged emphatically that new appointees to postdoctorates tenable overseas and paid from public funds should be admonished to perform scientifically as well as socially. Adams's point was heeded; he wrote,[23] "I was happy to hear that the National Science Foundation Fellowships have been set up with the idea that the American students will accomplish something in science while they are in Europe. Your restrictions sound most satisfactory." One is tempted to wonder if Adams's own scientific achievements in Germany in 1912–13 would have met his own standards of nearly half a century later. His European year had been nevertheless a key part of his education.

Adams was also an active member of the National Academy of Sciences from his election in 1929, serving both the academy and the National Research Council (its operating unit) in many ways. His trips to Germany and Japan in 1946–49 resulted from requests by the president of NAS, and his activities in international science were reflected in his chairmanship of the Advisory Committee on International Science Policy of the National Academy. This committee recommended strengthening the international science policy of the United States as a means of strengthening the free world by appointments of a highly qualified scientist as advisor to the State Department and of scientific attachés at major embassies abroad. This was done; one of Adams's Ph.D. students, W. R. Brode, was science advisor to the State Department in 1958–60. However, the overall impact of the science advisor to the State Department and of the overseas science embassies was not striking.

Adams served as foreign secretary of NAS from 1950 to 1954 and also headed the Research Council's Division of International Relations.[24] Correspondence indicates that international science was not then a major concern of the academy, compared to its later role with Harrison Brown as foreign secretary in 1962.[25] Adams also served on the investment committee of NAS for some years in the 1950s, but he resigned, saying bluntly that the academy had lost "literally millions" by having incompetent financial advisors.[26]

In 1962–63 Adams chaired a committee to review the desirable size of the NAS; in general the older members favored keeping it smaller and the younger members wanted it enlarged. Adams opposed the admission of members in the fields of social and behavioral sciences, but this was gradually done for the more "quantified" fields in the social sciences.[27] In his later years Adams kept letters flying to the academy president (F. Seitz, formerly a faculty member at Illinois, and later Philip Handler, an Illinois Ph.D. in biochemistry) commenting on investment and membership policies. One of his last letters (May 25, 1971) was to J. D. Roberts, chairman of the chemistry section, objecting to changes in NAS membership policy.[27] Adams preferred the earlier days of the academy when the number of productive scientists in the country was smaller and when it was possible to know personally nearly all the members. The present rather unwieldy size of the academy of over 1000 members reflects Adams's concern about changes in its numbers and composition. He wrote detailed comments on some sections of the manuscript of Cochran's history of the academy.[26]

Adams served on the council of the academy, 1931–34 and 1959–62, and enjoyed the academy's spring meetings in Washington. High points were dinners arranged by Speed Marvel for members of the wartime NDRC group who lived in the Washington area and some of the organic chemists attending the NAS meeting. These dinners were gala but informal affairs, with Adams kidding everyone and being kidded in return.

His continued role in the American Chemical Society placed him on the board of directors again from 1941 to 1949 and as chairman from 1944 to 1949. This was the period of Adams's service in Germany and his two trips to Japan. It is not surprising that international science was emphasized during his term; the board authorized him to aid the resumption of publication of the German compendia *Beilstein* and *Gmelin*, and the society collected journals that were sent to Germany and other European countries to help rebuild scientific libraries.[28]

Other problems during these years were the increasing size of ACS meetings, the revision of the constitution and bylaws, and the continuous problem of managing and financing the society's journals, particularly *Chemical Abstracts*.[28] After his term as chairman was over, Adams remained active in the ACS and his opinion was frequently consulted; he advised on membership affairs for several years.[29]

In September 1944, the ACS was offered a gift of the Universal Oil Products Company by its owners, a group of oil companies. A trust agreement named the Guaranty Trust Company the operator of the company as trustee, and the society the beneficiary of the income, which was designated for advanced education and research in the petroleum field, as very broadly defined. Thomas Midgley, Jr., president and chairman of the board of the society, reported the settlement. Adams's exact role in the drafting of the trust agreement is unknown, but his presence in Washington with NDRC and his reputation as a leader in research suggests his influence.

Midgley died suddenly after the establishment of the trust, and Adams succeeded him as chairman of the board in December 1944. The income from the trust accumulated, and in 1950 a committee of the board of directors, consisting of Adams, E. H. Volwiler, and C. A. Thomas, arranged to sell UOP stock to the public, eventually realizing $70,000,000. Reinvested, this formed the assets of the trust whose assets do not belong to the society but whose income has aided many investigations and is administered by the society under the name of the Petroleum Research Fund.[30]

Another of Adams's time-consuming activities was his association with Battelle Memorial Institute, for whom he was trustee from 1953 until his death.[31] This is a nonprofit research institute (similar to the Mellon Institute) founded by Gordon Battelle in 1925 to do research in many fields. It developed C. H. Carlson's invention, the xerographic process, and when this was sold to the Haloid (later Xerox) Corp. of Rochester, N.Y., the institute accepted a large block of Xerox stock. Voluminous correspondence with Battelle scientists and administrators and responses to financial reports testify to Adams's enduring diligence and courtesy. As a member of the finance committee for the board of directors, he faithfully attended monthly meetings. Deep personal interest in the synthetic and pharmacological study of the cannabinoid marijuana compounds launched by Battelle activated his letters.[32] In 1967 on a world tour he stopped at Battelle in Seoul, Korea, and wrote appreciative letters for the courtesy shown him, and shortly before his death he visited the Battelle Seattle Research Center set up in 1964–66.[33] In 1970 when he was over eighty, Adams was appointed an associate director of Battelle without a vote; however, he continued to attend directors' meetings and to receive reports. His last trip was to Columbus for a Battelle directors' meeting in the late spring of 1971.

John A. Wheeler, a distinguished physicist at Princeton, recalled his personal association with Adams at Battelle when Zay Jeffries and Adams, with his "friendly but direct-to-the-point way," persuaded Wheeler to join the board of trustees in 1959[34]. Their telling argument—that "serving on the board is even more than an

Illinois University Archives

Board members of Battelle Memorial Institute (ca. 1965): Zay Jeffries, Adams, John A. Wheeler.

Illinois University Archives

Board of Scientific Advisors to the Welch Foundation (1962): Adams, A. C. Cope, P. W. J. Debye, Henry Eyring, C. G. King, G. T. Seaborg.

education in science and technology. You have responsibilities for how the endowment is invested. Most of all, you get a feeling for what research means to the world"—spells out all the fascination that trusteeship held for Adams. He savored the financial problems and poured scorn on "highly acclaimed" financial advisors: "These people don't know what they're talking about, . . . we've got to . . . dump the dogs and get back on the band wagon." He participated in "policy decisions vital to Battelle's continuing success," questions involving xerography, the plutonium project in Washington State, and men to undertake the programs. "Men and policies were meat and drink to Adams."

Adams himself wrote in 1968, "My association with Battelle has been one of my favorite and satisfying activities. I have seen the organization grow and prosper, and my affection for it has grown similarly."[35]

The Natural Resources and Conservation Board (Illinois State Geological, Natural History and Water Surveys) counted Adams and other heads (Palmer, Noyes) of the Illinois chemistry department among its members; Adams served from 1942 until his death.[36] The group met around the state to examine installations related to its work. Between 1966 and January 1971, he attended thirteen meetings out of fifteen, some of which lasted two days. In 1969 he chaired an ad hoc committee to prepare legislation to transfer "to the board of trustees of the University of Illinois jurisdiction over the state Natural History Survey, the State Geological Survey and the State Water Survey," though it was never fully implemented.

Adams, as he wrote a member of the state administration task force, felt there were "no other state surveys in the country that can compare with them in quality and accomplishment." He felt this way, no doubt, because the qualifications required for appointment to the staff of the surveys were those required for the University of Illinois faculty. "It would be impossible to hire high caliber scientists for the Surveys if it was necessary to follow the procedures required for the line agencies in Springfield. The State Department of Personnel operates in quite a different way from the University. I stress this matter since it is so important to the effectiveness of scientific research organizations."[37] Adams's views on the indispensability of able people never changed.

Connection with the surveys led him into many matters. For instance, he wrote a member of the Natural History Survey, "Many thanks for sending me the book *Biology of the Striped Skunk*. . . .Frankly, I doubt whether I shall read it or not, unless I happen to have some intimate contact with a skunk which may induce me to learn more about him. Under any conditions, it will be a nice addition to my library." A more satisfying perquisite of the state service was a reserved parking space near the stadium and two tickets to all the football games; he enjoyed this advantage immensely.

Adams was also associated with the Robert A. Welch Foundation, which was established to support chemical research in Texas. Its scientific advisory board was a highly distinguished group of chemists with whom Adams served from 1955 until his death. He attempted to resign on May 19, 1971, but his resignation was refused on June 17, when he was in the hospital in his final illness. He proposed in his last letter to Welch on May 21, 1971, that the Welch Foundation offer a large money prize of $80,000 or more for outstanding research, and the foundation adopted this suggestion after his death.

Among his colleagues on the Welch board were Peter Debye, W. O. Baker, A. C. Cope, W. M. Stanley, Henry Eyring, and G. T. Seaborg. After Cope's death in 1966, Adams nominated E. J. Corey to replace him; Corey had begun a brilliant career of research at Illinois after his Ph.D. at MIT, and he moved to Harvard in 1959.

As always, Adams tackled the work of the foundation with vigor, reviewing research proposals, getting well-known overseas chemists to give series of lectures in Texas, considering appointments to Welch professorships of chemistry in Texas universities, advising about patents, and helping to plan the annual Welch Foundation Conferences on Chemical Research, which became a most useful feature of the foundation activity. This connection kept him in touch with current chemical research in many fields to some degree and, what was probably more important to him in his last years, allowed him to gather with old friends and associates at the meetings of the board.[38]

Adams organized the conference for 1968 on the topic "Organic Synthesis," selecting the speakers and discussion leaders and overseeing the details with great care.[39] In his opening remarks Adams discussed the basis for the choice of subject:[40]

> Most of the organic chemistry conferences that have been held in this country and abroad during the last fifteen years have stressed primarily mechanism, stereochemistry, confirmational [sic] analysis, or the chemistry of natural products. . . .It seemed timely, therefore, that at this Conference a little more attention might be paid to discussion of the synthetic aspects of the researches. . . the investigator who is familiar with the skillful in the art of organic chemistry will be able to realize results much more efficiently and more rapidly than the one who does not have that background. Whether the ultimate objective of the research is one in mechanism, in stereochemistry or in natural products is incidental.

The conference proceedings were an outstanding contribution to the literature of organic chemistry.

At the Tenth Conference on Chemical Research held in Houston in November 1966, Adams spoke in memory of Arthur C. Cope (1909–66).[41] Like Adams, an outstanding chemist and a very hard worker, Cope was almost Adams's heir apparent, serving on *Organic Syntheses*, *Organic Reactions*, and on many advisory boards with Adams, such as Battelle, Sloan–Kettering, and the American Chemical Society board. John D. Roberts, who was closely associated with both men, said of the 1946 period:[42] "The rising power in organic chemistry was Arthur C. Cope, a man chosen by Adams to be his successor in many things."

Adams recalled that during the past twenty-five years "in spite of the difference in our ages, Art Cope was a close professional and personal friend." After he gained a teaching position at Bryn Mawr, "our paths happened to cross several times in the next three or four years, and I became very much impressed with Art Cope and his potential." Cope moved to Columbia in 1940, with a recommendation from Adams to H. C. Urey, head of chemistry at Columbia, and because of Cope's outstanding record in wartime research, Adams suggested him to K. T. Compton as head of the MIT chemistry department. Here again he was impressive during the next fifteen years; "He had built up an extremely competent staff of younger men" so that the department "was rated among the most distinguished top few in the country."

Cope's own research was equally fine. "He discovered two reactions in organic chemistry which have been named after him. His work on the synthesis of mono-, di-, tri-, and tetra-enes from cyclooctane opened a field of tremendous interest to organic chemists," as did his recent exciting work on organic platinum complexes.

Cope "served as a director of the American Chemical Society longer than any other individual" and was chairman of the "finance committee over a long and critical period of time." Cope's other activities on "many important government and civilian committees around the country" met Adams's own exacting standards.

During the last year of Cope's life, his relations with Adams were not as close as they had been, due in part to differences of opinion over investment policies for institutions with which they were both connected.[42]

In the early 1960s, Adams reduced his consulting activities with Abbott and Du Pont, but he served a term on the board of directors of Abbott, 1952–59; he consulted for Coca-Cola into the 1960s. Although not officially consulting for Abbott, he continued to visit their laboratories fairly frequently in the late 1960s to see the Volwilers and to chat with the research chemists, some of whom had worked with him and all of whom enjoyed seeing him.[43]

Adams's travels and schedule are indicated in the following letter of 1960, showing how tightly his engagements were meshed,[44] even after his nominal retirement and when he was seventy-one:

> About the only time that I could attend a meeting during the three weeks starting March 23 is between now and the evening of April 8. My schedule for the following month is given below. I must keep open April 18 for a visit from the Deputy Polish Ambassador since I wrote to him that I would be here on that date and have not yet had a reply from him. On Thursday, April 14, I might be available to come to Chicago as I could cut short my attendance at the American Chemical Society meeting after noon on April 13. Unfortunately I have to be in Columbus at 9:30 a.m. on April 15.

March 23–April 8 (incl.)	Urbana
April 9–April 14	Cleveland
April 15	Columbus
April 16–April 19	Urbana
(tentative engagement on April 18)	Philadelphia and Washington
April 19–April 29	Urbana
April 29–May 11	
(May 4 and May 9 are closed dates)	

In considering Adams's activities during his whole career, one must keep in mind the difficulties in traveling from and to Urbana in those years. These are described in detail in a letter to Alfred P. Sloan of 1954.[45]

> I have not answered your most recent letter with your kind invitation to become a member of the Board of Trustees and the Committee on Scientific Policy of the Sloan–Kettering Institute for Cancer Research. I am very much interested but frankly question the advisibility of accepting owing to the fact that four additional trips to New York would be out of the question. My present obligations involve so much travelling that I hesitate to add others which require my absence from Urbana: If I lived in Chicago it would be simple but in Urbana it is different. Our rail connections with Chicago and the East are poor. It is necessary for me to leave here at 7:30 A.M. to reach New York the next morning by train. Usually I go to Chicago on a morning train and fly to New York reaching there about 5:00 P.M. This kills a working day. The Ozark Airlines from Champaign to Chicago has four flights a day. If this line were reliable, I could leave Urbana around noon or even at four P.M. and get to New York that evening. The planes, however, are so often late that one cannot rely on connections in Chicago unless one has two or three hours to spare. If the Chicago connection is missed, it is usually impossible to pick up passage to New York on a plane leaving shortly thereafter.
>
> When I travel East I try to do so at a time which will permit me to include several activities. Have the Board meetings for this coming year been set? If they have been I might be able to determine if it is feasible for me to accept your invitation.

Adams's evaluation of even this practical dilemma demonstrates the energy and will with which he carried out his responsibilities and pursued the aims of his career all his life.

Literature Cited

1. RAA, 25 and 26.
2. *Organic Syntheses* records, loaned by W. E. Noland.
3. RA to Norman Rabjohn, September 27, 1950, Ref. 2.
4. Norman Rabjohn, private communication, March 13, 1978.

5. Roberts started the practice of writing a signed preface describing the objectives of *Organic Syntheses* and the significance of the preparations in the volume at hand.
6. RA to Henry Baumgarten (Nebraska), then secretary, October 11, 1965.
7. Hundreds of documents are in RAA, 25 and 26; J. D. Roberts, private communication.
8. *Organic Reactions*, Vol. 25, Wiley, New York, 1977, has a historical account of the series by Blatt, pp. ix–xxiii; some early correspondence is in *Organic Syntheses* papers, Ref. 2.
9. Many documents about Adams and *Organic Reactions* are in RAA, 24 and 25; Adams's care as a proofreader was proverbial with the other editors.
10. Compton to Sloan, June 8, 1954; copy in RAA, 26, Sloan Fund.
11. Paul F. Douglass, *Six Upon the World*, Little, Brown, Boston, 1954, p. 130, calls Sloan naturally shy. The letters are all in RAA, Ref. 10; the dates range from July 6–16, 1954, just before the Adams retirement symposium.
12. Shown by a letter from Sloan to RA of August 10, 1954, addressed to Greensboro, Vt.; Sloan to RA, August 31, 1954, referring to two letters from RA of August 20 and 25. Apparently RA grew more enthusiastic about the potential benefits to research from the Sloan Foundation the more he thought about it.
13. A vivid account of Kelly by J. R. Pierce is in *Biograph. Mem. Nat. Acad. Scis.*, *46*, 190 (1975); it is clear that the Sloan initiative would receive his enthusiastic support.
14. RA's attention had been attracted to King by his article "A Foundation to Facilitate the Private Support of Pure Science," *Chem. Eng. News*, *28*, 1806 (1950), which RA had shown to Sloan at their first meeting.
15. Drafts and correspondence in RAA, Ref. 10.
16. A. P. Sloan to RA, April 20, 1955; J. R. and S. Cole, *Social Stratification in Science*, University of Chicago Press, 1973, p. 270.
17. RAA, 15, Alfred P. Sloan Foundation, Activity Report and Financial Statement, 1958–67; list of RA postdoctorates.
18. RAA, 48, Sloan–Kettering Institute; clearly many other papers have not been preserved.
19. List of RA's awards and offices: N. J. Leonard, *J. Am. Chem. Soc.*, *91*, a–d (1969); also unpublished list.
20. V. Bush, *Science, The Endless Frontier*, Government Printing Office, Washington, D.C., 1945.
21. The story of the National Science Foundation legislation from the standpoint of the National Academy of Sciences is in R. G. Cochrane, *The National Academy of Sciences: The First Hundred Years*, National Academy of Sciences, Washington, D.C., 1979, pp. 433–83; also V. Bush, *Pieces of the Action*, Morrow, New York, 1970, pp. 63 ff; D. J. Kevles, *The Physicists*, Knopf, New York, 1977, pp. 349–66; Don K. Price, *The Scientific Estate*, Harvard University Press, Cambridge, 1965, pp. 239 ff.
22. The surviving correspondence of RA with NSF is from 1958 to 1960; the letters mentioned are from RA to R. R. Brode, September 24, 1958, and to Vernice Anderson, secretary of the board, April 18, 1960.
23. RA to A. T. Waterman, director of NSF, February 16, 26, 1959; the second letter expressed RA's satisfaction.
24. Cochrane, Ref. 21, pp. 511–14.
25. RA–Atwood correspondence is in NAS archives, Administration: International Relations and Division of the NRC: International Relations.

26. RAA, 42, NAS, 1970–1; RA to Philip Handler, March 10, 1970; RA expressed similar views at other times in more caustic terms.
27. RAA, 42, National Academy of Sciences.
28. *A Century of Chemistry*, K. M. Reese, ed., American Chemical Society, Washington, D.C., 1976, pp. 33–6.
29. RAA, 29, American Chemical Society, 1966–71 and following folders.
30. *A Century of Chemistry*, Ref. 28, pp. 34, 73, 193; Midgley's report on the Universal Oil Products matter and the text of the trust agreement are in *Chem. Eng. News*, *22*, 1892, 1915 (1944); Midgley's death, ibid., p. 1896; RA elected December 2, 1944, as chairman of the board, ibid., p. 2162; legal matters connected with sale of UOP stock, *A Century of Chemistry*, p. 35; Ralph Connor to DST, May 18, 1979; information about the 1950 committee, private communication from E. H. Volwiler.
31. RAA, 17, 18, 19, 20, 30, 31; the extant RA correspondence and papers from Battelle cover the period 1962–71.
32. RAA, 19, BMI; E. B. Truitt, Jr., of Battelle to RA, January 31, 1969.
33. RAA, 30, Battelle Memorial Institute; RA to K. B. Hobbs, April 12, 1971; compliments Hobbs on a well-organized successful meeting in Seattle; on Battelle in Korea, correspondence is in RAA, 51, World Tour, 1967.
34. Private communication, J. A. Wheeler to R. M. Joyce, June 19, 1980.
35. RAA, 19, BMI, 1967; RA to B. D. Thomas of Battelle, May 24, 1968.
36. RAA, 24, 43, 44.
37. RAA, 44, II, Bd. Nat. Res. and Conser; RA to W. L. Blaser, December 9, 1968; this contains the striped skunk letter.
38. RAA, 27, 51; RA and the Welch Foundation.
39. RAA, 51, Welch Symposium on Organic Synthesis, 1967–68.
40. *Proceedings of the Robert A. Welch Conference on Chemical Research XII. Organic Synthesis*, Houston, 1969, p. 3.
41. Manuscript, RAA, 51, Welch; published in *Proceedings of Welch Foundation, Conference X*, Houston, 1967, pp. 1–3.
42. J. D. Roberts in *P. D. and the Bartlett Group*, Fort Worth, 1975, p. 368. RAA contains numerous letters from Cope to Adams. Unfortunately for the student of Adams, Cope's correspondence with Roger Adams was not retained in Cope's papers at MIT (G. A. Berchtold to DST, March 17, 1978).
43. The winding down of Adams's consulting activity at Du Pont is discussed in RAA, 33, Du Pont Co.; RA to R. M. Joyce, January 23, 1964; with Abbott, RAA, 27, Abbott, RA to A. W. Weston, March 13, 1967.
44. RAA, 26, Sloan Fund; RA to Warren Weaver, March 23, 1960.
45. RA to A. P. Sloan, October 9, 1954; Ref. 44. This is the one of the few mentions by Adams of the problems of travel from Urbana; he probably spelled it out for Sloan because of the latter's intention to give valuable support for basic research.

Overseas Travel, 1949–70

Trips to foreign countries for scientific meetings or for pleasure were frequent for Adams after 1950. Attendance at international symposia, particularly those organized by the International Union of Pure and Applied Chemistry (IUPAC), of which he was an officer, kept him in touch with his overseas colleagues and increased the ties between American and overseas science, ties that led to an extraordinary amount of personal correspondence between Adams and his friends and admirers in many countries.[1] The breadth and cordiality of this correspondence is striking; Adams's unfailing warmth and pleasure in personal contacts and in expanding science make a lasting impression. He made seventeen trips abroad from 1953 to 1970, in addition to several that were family outings and that are described separately.

In April 1953 he attended scientific meetings in Spain, described in detail in letters to Mrs. Adams. Photographs from his trip show Adams with his friend A. R. Todd, the distinguished British chemist, and Spanish colleagues.[2] As usual on his travels, Adams absorbed an amazing amount of information about the country visited and was able to give a comprehensive account of the scientific and economic characteristics of Spain.[3]

Adams and Mrs. Adams attended the 18th IUPAC Conference in Zurich in August 1955, following it with a trip to Germany, Austria, and Yugoslavia.[4] In a letter of September 21, 1955, to a Japanese friend, he reflected:

> ...We had a good opportunity to see how a communistic system functions. I can assure you that the difference between the attitude of people in Yugoslavia and any of the adjacent non-communist countries is very marked. There seemed to be no energy or ambition among either the laboring class or the intellectuals, since there is no incentive to induce an individual to do his best work.

Later Adams appeared to mitigate his view of communism in Yugoslavia as the inroads of capitalism brought some progress.

The Du Pont Company wished to learn firsthand about original young chemists overseas in order to make grants to support promising research programs and to invite some of the chemists to lecture in this country. Adams made a carefully prepared trip to Germany in 1958, where he visited many university laboratories, sought opinions about young chemists from the older established chemists, and picked out the most promising based on his interviews and the comments of others. As a result of his intensive survey, Du Pont invited several young German chemists to this country to present seminars at Du Pont and elsewhere.[5] In February 1962 he made a similar trip to Britain, where he visited and commented on fifteen universities in two weeks. Although his well-planned travels in Britain were simplified by having a car and driver, it was an impressive exhibition of stamina and sound judgment for a man of seventy-three.[6] After his 1964 trip to

Japan, he wrote a report to Du Pont on Japanese laboratories and organic chemists and mentioned particularly K. Nakanishi, later a professor at Columbia.[7]

In 1959 Adams traveled in Japan for several months with two scientists from Abbott Laboratories, A. W. Weston and J. C. Sylvester, to introduce them to Japanese scientists and to visit research laboratories.[8] This was apparently Adams's first visit to Japan since the 1948 mission, and he was greeted everywhere with great cordiality and deep respect by Japanese chemists, including numerous former students. Groups gathered sponaneously in the lobby of the Imperial Hotel as he came and went, just to see him and to welcome him to Japan. He greatly enjoyed the social arrangements and relished, in particular, playing the Japanese pinball machines, "Ochinko."[9]

In August 1960, funded by a travel grant of $1200 from the National Science Foundation, he traveled to Australia for an IUPAC chemcial symposium and visited universities in New Zealand before the Australian meeting.[10] This was followed by a week-long trip to New Guinea in a chartered DC-3 by a group of distinguished American and foreign organic chemists. Adams was the hardiest sightseer in the group, missing nothing and inquiring into everything, from early morning until late at night. His energy astonished his companions, all younger than he.[11] A consulting report on Australian and New Zealand chemistry went to Johnson and Johnson Company, where his former student, W. H. Lycan, was vice-president for research.[12] Because Adams had been unable to accompany Lycan to Japan in 1960, he also advised Lycan about Japanese chemistry, and the latter found that Adams's "name was still magic in Japan, even though it was more than twelve years after his second Scientific Advisory Mission to Japan."[13]

Adams had a great love for and an interest in jewels, and on many of his trips abroad, especially after World War II, he bought unset stones that he had mounted after returning home. He relied upon his many foreign acquaintances to guide him in the special gems of the area. On his trip to the Orient in 1960, for example, he purchased unset Australian opals with the advice of Francis Lions at the University of Sydney, and in Tokyo in 1959 he was an honored buyer of cultured pearls from K. Mikimoto, Inc. Mikimoto's grandson, Hiroshi Yokohama, had entered the United States in the fifties with Roger as a sponsor and taken a degree at Illinois in electrical engineering. This Japanese family remained the Adams's devoted friends for years, and the Yokohamas and their four children came to Urbana for Roger's eightieth birthday party.[14] In addition to his desire to please his wife and daughter, the purchase of jewels satisfied two deep-seated traits, his love of a good investment and a more deeply hidden aesthetic sensitivity.

In 1962 Adams toured the Old and the New Worlds, his Du Pont trip to England in February and his trip to Prague for the IUPAC meeting in August. He furnished the NSF, his trip sponsor, with detailed comments on the meeting and situation of science and scientists in Czechoslovakia.[15] With hardly a break he was off to Argentina in September for the Eighth Latin American Congress of Chemistry where he talked on "Fifty Years of American Chemistry, 1912–1962," and

"Since I was going that far, I added three extra weeks and spent them partly in the Andes in Bariloche and environs and partly in Brazil where I visited the Universities in Sao Paulo and Rio."[15]

To a Czech chemist, for whom Adams found a postdoctorate at Illinois in 1963, he wrote on October 23, 1962:[16]

> ...I have just returned from a most interesting trip to South America. I was in Buenos Aires during the revolution so had a chance to see how they operate one of those. The curious thing is that the public is not at all involved, it is merely the battling of two generals using the soldiers [that] have been assigned to them for their own benefit.

He mentions that he visited Brazilia, as well as Sao Paulo and Rio, and after his return he corresponded with several Brazilian chemists, particularly his old friend V. Deulofeu (who had been on the New Guinea jaunt), about a visiting American professor of chemistry for Brazil.[17]

A letter to D. M. McQueen of November 19, 1962, describes some photographs of a coffee plantation in Brazil.[18]

> The tall trees at the left are Leucaena glauca and the coffee trees are on the right. The rows of coffee trees and shade trees are alternated. I expect that by this method the flowers and beans develop more gradually and presumably a better crop results. When the coffee trees are small Crotalaria bushes are planted beside them to shade them much more than the Leucaena glauca. When the coffee trees are three or four feet high these trees are pulled out and the Leucaena glauca take over their job of shading.

To Adams these were more than just coffee trees and shade trees; the *Leucaena* tree was the source of the alkaloid leucenol, whose structure he had established, and the *Crotalaria* trees furnished members of the *Senecio* alkaloid series.

July 1963 found Adams in London for an international congress; he had stopped in Boston on the way to see his sister Emily, who was well and cheerful. The London visit was a festive occasion; he met hundreds of people and relished the constant entertainment. At the congress banquet of 800 quests, Adams sat at a front table with Sir Robert and Lady Robinson of Oxford. H. Erdtman of Stockholm, who spoke for the guests, referred to Adams as the "Robert Robinson of America," a high compliment indeed, as the scientific world regarded Sir Robert as the international master of organic chemistry for his generation. "Such advertising doesn't do any harm," wrote Adams. On the lighter side, Adams and George Mercer, his associate from NDRC and German days, went to the racetrack in London and lost three shillings on bets.[19]

On July 19 he flew from London to Warsaw for IUPAC, where he received a gala welcome from T. Urbanski of the Warsaw Technical Institute and from the Polish Academy of Sciences of which Adams was an honorary member. He went sight-seeing, had "two wonderful days in the mountains," visited Cracow, characteristically tried to help a Pole get an American visa, and returned via Copenhagen and Frankfurt.

In the summer of 1964 Adams returned to Japan for a IUPAC International Symposium in Natural Products. During this trip he had a severe attack of asthma (clearly not a heart attack), described thus by Nakanishi:[20]

> It was probably in 1964 just before the IUPAC International Symposium on the Chemistry of Natural Products, April 12–18. He was visiting the Tohoku University in Sendai, which is where I used to be just before joining Columbia University.... After the lecture and the usual dinner reception, we saw him off at the hotel. Shortly later, I received a phone call from the hotel manager that Professor Adams had collapsed in the hotel lobby. We immediately got things arranged and assigned to him several students who in turn looked after Professor Adams all night long. It was a very mild stroke and being an energetic man as he was, he soon started chatting and telling the students many interesting episodes, etc. about chemistry and life. The next morning I received raving reports from the students about the great Professor Adams.

The account of Adams discoursing to an admiring group of students while he recovered from his indisposition is diverting and completely in character.

Adams's journey to Moscow in July 1965, also to attend an IUPAC meeting, is brought to life by a firsthand account.[21] E. H. Volwiler and his wife met Adams at Berlin and the three flew first to Vienna, then with great difficulty made the 175 miles to Budapest in a Viennese taxi, surmounting flooded roads, a ruined tire, and disagreeable Hungarian officials at the border. At Bucharest, their next stop, Adams was not feeling well at the airport, but with minor treatment (ephedrine) at Kiev, they made it to Moscow. Volwiler's tale continues:

> ... At about ten P.M. ... down the stairs came Roger, with an orderly on each arm. He wryly explained to us that he had called the hotel doctor, who at once ordered that Roger be taken to the Botkin Hospital.
>
> Two Americans standing at the desk noted our concern, and asked to be of help. They were Samuel Jaffe, who had been the ABC news correspondent in Moscow for four years, and spoke Russian. (Shortly after, Jaffe was expelled from Russia for his reporting.) The second man was Peter Jennings, just on his way, for ABC, to Viet Nam. These men were acquainted with the American Embassy staff, and they at once got busy with the staff, including the American doctor.
>
> The next morning, we talked to the doctor and others at the Embassy. The doctor said he would check with the staff at the hospital, but would have to be delicate about offering suggestions. He said that ordinarily, when Americans become ill in Russia, the Embassy prefers to send them home at once; if they cannot travel that far comfortably, they can be sent to the American Hospital in Frankfurt.

In the afternoon (Monday), Dr. Bazzell stated that Roger was suffering from severe pleurisy with some emphysema, and, on Dr. Brazzell's suggestion, had been put on antibiotics (chloramphenicol).

On Tuesday afternoon, the hotel clerk gave me a slip of paper bearing the name of the Botkin Hospital (in Russian of course), which I handed to the cab driver. He deposited me at the corner of the hospital, near the entrance. It is a very large structure, covering a block and three stories high. It is built of yellow brick, with marble stairways (broken in places). There was no admissions desk there, and I saw no elevator. Now I started on my search for Roger, in a thousand bed hospital. My only clue was Dr. Brazzell's remark that Roger was on the third floor. So up I went, and nobody asked me a question. And in the second patient room that I entered, there was Roger in the first bed! He astonished me by his good appearance and heartiness, though at the moment he was protesting the yogurt which the nurse was urging on him; finally she gave up, with a good-natured touseling of his hair.

The room, apparently more or less standard in the hospital, was long and narrow, with one window at the end. There were three cots, end for end, with apparently clean but dingy-looking bedding.

Roger said that it had taken the woman doctors a while to be convinced that he was not suffering from a heart attack. They had given him two EKG's, and X-rays. He was anxious to get out, but had not succeeded so far. He invited me to take a little walk with him, especially to see the men's toilet, across the hall. That was quite a revelation. The one window had a quarter-inch mesh screen, full of soot and evidently untouched in years. There was a row of men's urinals, but apparently out of order, because all were covered and marked. For sitting down, there was a row of stalls, without doors. There was no paper—a patient was supposed to bring his own newspaper. The bowl was porcelain. It had a wooden seat, which stood alongside to be picked up and placed on the porcelain when needed.

I left Roger, both of us hoping that reason would soon prevail so that he could be released to attend some of the Congress sessions. The Russian scientists running the Congress knew of Roger and his illness. On Wednesday, they sent a young English-speaking Russian physicist in a Government car to take Roger from the hospital back to the National Hotel. Roger protested to the physicist about being held so long after he was well enough to be released. "Well," the man replied, "you must understand our system of medical care—it is different from yours. You see, each hospital has a certain quota of personnel—doctors, nurses, pharmacists, janitors. They are all busy and fully occupied, and they are supposed to stay that way. So, until there is a new patient to take your place in a hospital bed, you must stay." Incidently, Roger was not charged anything for his medical episode.

He had dinner with a group of his friends that evening, and attended the closing session of the Congress the next day. He cancelled his plan to go on a post-Congress tour, to Tbilisi, and flew home alone. They refused to refund any of his Tbilisi tour money.

For three months in 1966, January to March, Adams toured Africa with the

Volwilers.[22] The tour was "entirely successful," "eye opening"; he talked and wrote about it in detail, he observed and photographed the landscapes and animals, he commented with his usual acuteness on the economy, politics, and leadership of the countries he visited, and he delighted in presenting slide shows to his audiences in Urbana and elsewhere.

His detailed narrative to his daughter in a long letter of March 13, 1969,[22] shows that he found Kenya, Tanzania, and Uganda particularly interesting to visit; "Treetops" in Nairobi was "a unique place." He recommended the Ngorongoro Crater in Tanzania and its animals; he liked Rhodesia, Salisbury, and "of course, you would then go to Victoria Falls for a couple of days. Personally I was disappointed in Victoria Falls because I had anticipated something different from what I actually saw. The falls are tremendous; but it is impossible except from the air to get a good view of all the Falls, which are circular and about a mile long, because the water pours into a ravine and you can't get a view from the bottom of the falls." In a lecture of 1966,[23] Adams discussed in detail his disgust at shooting safaris, his experience in photographing wild animals, Rhodesia and the Ian Smith government, "Apartheid" in South Africa, his six days in Uganda, and his impressions of many other African countries. It is clear that Adams did not read up beforehand on the countries he was visiting, though he may have questioned people who had seen them. However, there was little that escaped him on his travels, and his natural ability as a quick student enabled him to absorb and organize a large body of information and impressions.

In Stockholm for the IUPAC Natural Products meeting in June and July 1966, he substituted for A. C. Cope in introducing Professor Karlson from Marburg. As usual on his European trips, he visited the Battelle Laboratories, one in Frankfurt and one in Geneva.[24]

From January to March 1967 Adams went alone on a round-the-world tour, described in great detail in a diary apparently typed from his letters after his return and intended probably for his daughter and his sister Emily.[25] The diary shows him making friends rapidly, particularly with a group of four from Nashville, being entertained by scientific colleagues in many places, and viewing all the sights and people with an unjaded, inquisitive eye.

He flew to Athens, where he was more interested in gold jewelry than in the statuary, took a vividly described trip to Corinth, visited Istanbul for two days, then went to Beirut and Palestine. Like many tourists, he doubted the authenticity of some of the sights in Jerusalem, but he said he was "having a good time." In Egypt he found the "King Tut relics fabulous." Bridge and gin rummy with his Tennessee friends enlivened the boat trip to Karachi and Bombay from Cairo.

In India he was lavishly entertained by two former students, Govindachari and Nair. The former was head of a Bombay pharmaceutical laboratory owned by the Swiss firm Ciba, and Adams took great satisfaction in the professional success of these students. On tour again, the Taj Mahal was "stupendous," and learning that Shahjahan's wife, in whose memory it was built, had died with her thirteenth child at the age of 37, Adams commented, "She must have led a very pregnant life."

After seeing Calcutta and Delhi, he flew to Bangkok, Singapore, and Hong Kong, where he became anxious to get home, yet he spent several days in Seoul, Korea, principally at the Battelle Institute there, where he was "stuck to talk," and he was much impressed with the institute.

In Tokyo he joined the members of his tour again and visited Hiroshi Yokohama, but because he had caught cold in Singapore, he did not attempt to see all his friends in Tokyo. Tokyo to Hawaii was a "terrible night flight," but he revived and, undaunted, took a tour of the islands; although he had done it before, he was pleased to repeat it.

He was glad to arrive back in Urbana, but he evidently enjoyed the trip, even though he had numerous complaints for the travel agent when he returned. His desire for travel had been temporarily satisfied; he wrote just after his return to Urbana, "At present I have no hankering to travel any more since this last trip kept me on the go and part of the time I wasn't feeling too well. I doubt very much that I will want to visit Prague again for I was there in 1964, and Lucile and I had a month in Jugoslavia about seven or eight years ago.... I shall be glad to relate some of my experiences on this last trip."[26]

In May 1968 he attended a IUPAC Natural Products meeting in London and again visited the European Battelle Laboratories. His next IUPAC meeting was in Mexico City in April 1969, when he was eighty. He wrote to Carl Djerassi for information about interesting trips around Mexico City and spent several days in Yucatan, "most interesting days," but too hot. As usual after an international meeting, he wrote letters to his hosts and organizers of the meetings, complimenting them on their program.[24]

Adams's last trip to an IUPAC meeting was in 1970, to Riga, for a Symposium on Natural Products. His daughter accompanied him and the trip was far happier than his earlier visit to Moscow. Adams wrote Henry Gilman,[27] "I liked Riga. Although Lucile and I had rooms reserved in the Riga hotel, Professor Shemyakin moved us to a hotel near the beach in a very delightful location. It would have been very inconvenient for the meetings 18 miles away if he had not provided me with a private car and chauffeur for the week we spent in Riga.... The meeting was unusually successful considering that the hotel accommodations were entirely inadequate. There were 1,500 attendants, of whom about 1,000 were Russians." This was his last trip out of the country.

Literature Cited

1. Of this voluminous correspondence, hundreds of letters are preserved; RAA, 13, 33–36, 39, and elsewhere.
2. RAA, 63, Group Photographs, 1935–53.

3. RAA, 40, Midwest Award; address by RA, "A Glimpse of European Science"; RAA, 6, Spain, letters to Mrs. Adams.
4. RAA, 34, P, correspondence with M. Prostenik; 38, IUPAC; RA to R. Yokohama, September 21, 1955; RAA, 39, Japanese Correspondence. The official IUPAC American delegation consisted of F. D. Rossini, chairman, Ralph Connor, vice-chairman, P. D. Bartlett, W. R. Brode, H. E. Carter, and A. C. Cope.
5. RAA, 21, 22, Du Pont; T. L. Cairns to DST, May 9, 1979.
6. RAA, 33, Du Pont European Trip, England (2/62), gives RA's schedule, notes on chemical research programs and British chemists, correspondence, and recommendations to Du Pont.
7. RAA, 13, Information for Du Pont on Japanese, 1959–64.
8. RAA, 39, Japanese Trip, contains a 155-page report by Weston and Sylvester to Abbott; also in RAA, 15, Abbott Correspondence.
9. A. W. Weston, private communication.
10. RAA, 38, IUPAC, 1945–69; RAA, 17, Australian Trip; RAA, 6, New Zealand, Australia, and New Guinea, containing letters to Mrs. Adams. The NSF travel grant is in RAA, 23, National Academy of Sciences, 1953–67.
11. N. J. Leonard, private communication.
12. RAA, 54, Reports by Adams, 1947–62.
13. W. H. Lycan to DST, June 12, 1979.
14. RAA, 35, Foreign Correspondence; RAA, 39, Japanese Correspondence; private communication from Lucile Adams Brink.
15. RAA, 43, National Science Foundation #1; RA to W. R. Kirner, October 24, 1962, ibid. Kirner was program director for chemistry of NSF, and Adams had received a travel grant from NSF to go to Prague.
16. RAA, 46, Trip to Prague; RA to Jiri Jonas; Jonas became a tenured staff member at Urbana.
17. RAA, 33, Deulofeu; RAA, 48, South American Trip.
18. RAA, 33, Du Pont Company, 1961–71. The South American trip is described in detail in letters to Mrs. Adams in RAA, 7, Argentina.
19. RAA, 7, England and Poland, 1963; details in letters to Mrs. Adams.
20. K. Nakanishi to DST, May 22, 1978.
21. E. H. Volwiler, private communication.
22. RAA, 28, African Tour.
23. RAA, 44, Lecture at Carl Noller Symposium, May 20, 1966; the last 20 pages describe RA's African trip.
24. The IUPAC meeting documents are in RAA, 38, IUPAC meetings.
25. RAA, 6, Diary of World Tour 1967.
26. RA to E. H. Volwiler, March 10, 1967.
27. RA to Henry Gilman, December 16, 1970; January 7, 1971. Other correspondence in RAA, 47, Trip to Riga.

The University, Urbana, and Informal Life

From the early 1920s, when he first emerged as a major figure on the local and national scene, and especially from 1926, as head of the largest and best-known department in the College of Arts and Sciences at Illinois, Adams was naturally called on for service on many university committees and groups. He managed somehow to sandwich such work into his schedule, partly to represent the interests of the chemistry department and partly because he regarded it as an obligation to the university as a whole. His effectiveness and sagacity in university affairs were so well known that several times he could have become president of Illinois or some other university if he had desired. We make no attempt to cover his activities in detail but choose some well-documented examples from his later years of service to the university and to the Urbana–Champaign community.

Around 1950, reports became current of a material called "Krebiozen," prepared by an undisclosed process by Marco and Stevan Dorovic and alleged to be an effective treatment for cancer. The material was championed strongly by A. C. Ivy, M. D., vice-president of the Illinois School of Medicine in Chicago, and a highly respected figure in the medical world. However, the refusal of its sponsors to allow controlled tests of Krebiozen, to describe its preparation, or to submit it to chemical and pharmacological study led to an extremely bitter controversy in which Ivy was opposed by the medical profession. G. D. Stoddard, then president of the university, was drawn into the dispute against Ivy and eventually he was sued for libel by Ivy. Adams served as chairman of an ad hoc committee of the university senate to investigate the Krebiozen matter, another member being his biochemist colleague, W. C. Rose. Adams's committee issued a report in December 1952, which was critical of Ivy; the state legislature carried out an investigation of its own. Adams obtained a statement from the distinguished University of Chicago physiologist Anton J. Carlson supporting the position of the Adams committee, and Stoddard praised the committee for its report. Eventually the libel suit was dropped.[1] Adams's willingness to investigate and report on a matter as involved and emotionally charged as Krebiozen showed his courage and the seriousness with which he regarded his professional responsibilities.

The retirement pay for University of Illinois faculty and employees was extremely small, especially for those who retired before 1959. Although Adams was well provided for personally through his consulting, investments, and Mrs. Adams's independent estate, he saw many of his retired colleagues in very straitened circumstances after many years of devoted service to the university. In 1961 Adams started an active campaign to improve retirement benefits; he organized a committee, collected data from other institutions, appeared before a committee of the Illinois trustees and the Illinois Budgetary Commission, and eventually got considerable improvement in retirement pay. Either in person or in writing, Adams was a tenacious and formidable advocate, equally adept at using the rapier

Illinois University Archives

Adams in his scientific library.

or the bludgeon, and his great personal prestige gave him additional force.[2] Furthermore, he was always in complete command of the facts.

Adams's judgment was valuable to a group called University Patents, Inc., formed in 1964 to take out and administer patents on potentially profitable research results from the University of Illinois. Obviously modeled after the Wisconsin Alumni Research Foundation, which administered the valuable Steenbock patents on irradiation of foods to produce vitamin D activity and the Link patents on blood anticoagulants, the group had extremely optimistic hopes of bringing additional funds to the university, some of which would be used for support of research. Adams's experience as consultant and as holder of numerous patents himself made him a key man on the board of directors of the group. He attended meetings conscientiously and tried to make the expectations of financial returns more realistic and less optimistic. It is not clear that any very profitable patents resulted from the venture.[3]

Although Adams was not in sympathy with the student mores and protest movements of the 1960s, he did appear in the Urbana court in 1967 as a witness for the defense of a student charged with possession of marijuana. Adams felt that marijuana was not properly classified as a narcotic, and the student was acquitted.[4] Adams refused to testify for the defense in other cases where the *sale* of marijuana was the charge.

In addition to Adams's membership in the Urbana Rotary Club, his most notable service to the local community was his membership on the board of directors of the Champaign National Bank. He was a valuable member of the board because of his acumen and wide experience in finance and also because he never wanted to temporize in making hard decisions, once the necessity for them was clear.[5] The bank honored Adams after his death by setting up a room that displayed his medals and awards in a special Roger Adams Exhibition.[6]

Adams invested in several local enterprises; one was Delmont Village, an apartment complex of which he was a director and for which he wrote detailed comments about the best type of apartment.[7] Another was the Urbana–Lincoln Hotel, in which investment had been urged at an earlier time because it was a "civic enterprise" benefiting the city of Urbana. In 1954 the Urbana Hotel Co. changed hands, and Adams launched a crusade to protect the minority stockholders, of whom he was one, whose rights he felt had been ignored. In this, however, he was unsuccessful in spite of a detailed study of its accounts, letters to the new management, and a protest published in the local newspaper.[8] As in the university retirement benefits, the money involved was not the point to Adams; he felt that he and others were being unfairly treated. As he made clear in correspondence relating to many of his overseas trips, he expected to get what he thought he was paying for, and he protested vigorously if this was not the case. Adams also invested in an Urbana beauty school, an enterprise apparently more productive of amusement to his friends than of financial returns to him.[9]

Despite these multifarious scientific, administrative, and public service obli-

Illinois University Archives

E. H. Volwiler, Mrs. Volwiler, Mrs. Adams and Roger (1950).

Illinois University Archives

With his grandchildren.

gations, which Adams took very seriously, Adams did have an informal life with family, friends, some hobbies, and recreation. His personality made many of his more formal, semipublic activities a source of recreation and relaxation to him, to a degree impossible to one who lacked his interest in people. Consulting visits, with their frequent accompanying poker games, board meetings with old friends, research conversations with students, and many of his other professional or public service activities thus were pleasant and were free of nervous tension. His ability to carry on determined or even heated arguments without personal rancor and to put aside a disagreement after it was over saved him from brooding about the past and from the corrosive effect of harboring vindictive feelings. Thus the qualities of personality that allowed him to do so many things protected him very largely from the strain and worry that might have led to exhaustion and physical collapse.

Adams was devoted to his wife and to their daughter Lucile. Lucile, Jr., on the advice of her old friend, the eminent biochemist Vincent du Vigneaud, majored in physiology at Mount Holyoke, where she graduated in 1948. Because she had seen little of her father since 1941 due to his war work and later journeys, she asked for a trip with her parents as a graduation gift, which took the form of a Caribbean cruise including a visit to Venezuela.[10] A year later they all toured Europe together, Roger giving papers in Switzerland and Amsterdam,[10, 11] and in 1951 they visited Scandinavia.[12] "The trip was restful, although 3 days were rough and Mrs. A. and my daughter were laid low. We stayed in Oslo a few days, motored down the west coast of Sweden then crossed to Denmark for 10 days. From there (Hirtstal in Northern Jutland) we ferried to Christiansand in Norway and thence to Bergen by car. Norway is certainly rugged and poor, and I don't see how they live. The people are rationed on clothes and many foods."

Lucile, Jr., worked in Boston at the Harvard School for Public Health in the field of nutrition before returning to Urbana, and in August 1952, on the anniversary of her father and mother's wedding day, she married William Ranz of the Illinois School of Engineering. Roger and Mrs. Adams were devoted grandparents to the four children, Beth, Christina, Roger Adams, and Jennifer.[10]

Mrs. Adams' health failed in 1963, and after she suffered a series of strokes at their summer home in Vermont, Adams had her flown to Urbana in an ambulance plane. She weakened rapidly and died in January 1964. Adams was severely affected by her death, as a poignant letter to one of his oldest friends, Henry Gilman, shows,[13] and his depression lasted for a long time. In a letter of 1965 Adams mentioned that it was "dreary" living alone,[14] although there was never any suggestion that he would leave his home and Urbana. His friends, the Blatts in New York and the Copes in Boston, arranged to have dinner with him when he was alone for the evening in either New York or Boston so that he would have congenial company if he wished it.[15]

His daughter and grandchildren were increasingly close to him, and he visited them frequently in Minneapolis. A month's tour with them to Alaska in the late 1960s was a particularly happy time. In Urbana he and his longtime colleague and friend, W. C. Rose, attended the ballet and symphony concerts together in their later years,[16] although music had never been one of Adams's active interests. He enjoyed playing bridge in 1968, as he had in the earlier days at Urbana,[17] although it is not known if he had a regular group. In spite of his family, his host of friends, and his many activities, Adams was lonely in his last years. His outgoing personality and his ability to make friends easily covered a hard core of privacy, which he did not often breach.

His surviving sister, Emily, was the object of his constant solicitude in his later years. In 1964 she suffered a stroke, and Roger placed her in a nursing home on Cape Cod where he paid her expenses, visited her when possible, and wrote her newsy letters about his travels and activities. His devotion is touching, as shown in letters to her and about her; with characteristic foresight he set up a trust fund to care for her if she outlived him. Emily died in 1970.[18]

One of Roger's first cousins in New Hampshire figured in a pleasant family episode. Adams went to the University of New Hampshire at Durham in 1966 to take part in the dedication of a new chemistry laboratory named after his friend Charles L. Parsons, who had taught at Durham before becoming secretary of the American Chemical Society. Ann H. Adams, Roger's first cousin, lived in Durham, and he visited her there for the first time in many decades. He later corresponded with her fairly frequently, and she wrote him about the early history of the New Hampshire branch of the Adams family.[19]

Adams continued to spend his summers, as the family had for many years, at the summer home or "camp" in Greensboro, Vermont, which had belonged to Mrs. Adams, and here he enjoyed the visits of his daughter and grandchildren and the companionship of old friends of the summer colony. He gardened, sailed, and played golf and cards occasionally. He still loved poker and hoped that the grandchildren would be good players; Jennifer, the youngest, showed the most skill. All of them enjoyed playing Russian bank, the children trying hard to beat Grandpa.[10]

Especially fond of native trees and ferns, he took pride in tending his huge bed of maidenhair fern (his favorites) and the many varieties of trees with which he restored his battered woods after the destructive hurricane of 1938.[10] He worked at odd jobs around the camp and carefully supervised its upkeep. His papers contain some interesting exchanges with local Vermont craftsmen, who evidently thought that all "summer people" would pay any sum charged without protest. They soon found that he was an exception.[20] Adams also looked after his Urbana home with great care, especially the gardens.

In 1965 Adams was advised by a neurologist to wear a corset to hold his back straight for prevention of sciatic pain. "This I did for a few hours in New York when the temperature was 95, and soon decided that the jacket was more uncomfortable than the sciatica.... Actually I found one day at camp [in Vermont] when I was lifting heavy rocks after having a ledge blasted that my back and leg were better the next morning. Apparently I twisted in the right direction. For the past week my leg has bothered me less than it has most of the Spring."[21]

Honors continued to come to Adams after 1945 for his teaching, research, and public service activities. He received honorary degrees from ten universities, including Illinois, Yale, Harvard, Michigan, Pennsylvania, and Rochester. Conant's citation for the Harvard degree said: "Roger Adams: Famed as a teacher and investigator, the war leader of American chemists; both men and molecules react benignly as he plans."[22] His Illinois degree in 1957 pleased him particularly; he wrote Karl Heumann, one of his Ph.D.'s, that Illinois degrees "are scarce as hens' teeth," and to Ralph Connor he said, "I do appreciate very much an honorary degree from the University of Illinois since they are rarely given to a faculty man. It will mean much more to me than the degrees I have received from some of the Ivy League universities."[23] He received the Franklin Medal of the Franklin Institute at a ceremony in Philadelphia in 1960; he wrote Henry Gilman on November 1, 1960, "Mrs. Adams and I enjoyed very much going to Philadelphia and attending the award dinner. It was quite impressive. The only difficulty with an award of this kind is that one is required to give a speech, and since it had to be popular in character I have been worried about it since late spring."

Adams received most of the medals given by the American Chemical Society and its local sections, such as the Willard Gibbs Medal (1936), the Charles Parsons Award (1958), the Priestley Medal (1946), the Theodore William Richards Medal (1946), and many more. The last award no doubt recalled his unhappy period of doing research with Richards in 1911. He was an honorary member of many academies and scientific organizations, both foreign and domestic.

His services in World War II were recognized by the American Medal for Merit (1948), by letters of commendation from President Truman, and by appointment as Honorary Commander of the Civil Division of the Most Excellent Order of the British Empire (C.B.E.) in 1948. President Johnson presented to him the National Medal of Science in a White House ceremony in 1966, which his daughter attended with him,[24] and in 1967 the state honored him with the Order of Lincoln for Science conferred by the Lincoln Academy of Illinois, an award that his daughter accepted for him in his absence. The list of his awards, honors, and offices covers three typed pages.[25]

Adams enjoyed these distinctions and his clear judgment made him realize that he deserved them. The same judgment, however, never allowed him to confuse the substantial with the unsubstantial. To him the substantial was the excitement of good research, the training and subsequent growth of his research students, the reputation of the University of Illinois, the solid accomplishment of his public service activities, and the pleasures afforded by his friends and family. In his hierarchy of values, recognition and honors were pleasant and had some importance, but they were not the motivation for his hard work. He took particular personal satisfaction in his scientific work on gossypol and marijuana.[26]

Adams's shrewd appreciation of the usefulness of external honors to the Illinois department is shown in this letter to a departmental aide in 1970:[27]

Illinois University Archives

Receiving an honorary degree at Illinois (his twelfth) from President David Henry (1957).

Lucile Adams Brink

Adams receiving the National Medal of Science from President Johnson, 1964. Center, D. F. Hornig, Special Assistant for Science and Technology to the President.

> I am sending you the pamphlet presented to each one who attended the President's Dinner at the American Chemical Society meeting in Chicago. Included is a list of the past presidents of the Society. It seems to me that here is an opportune time for a little advertising in your next annual report of the department activities. You will note the unusual record of the University of Illinois. Of the 82 past presidents, including Byron Riegel who was President during 1970, there are 14 who have been either students or staff members at the U. of I. I indicated an additional two with question marks. Bartow was a professor here for quite a number of years prior to 1920 when he left to become head of the Department at the University of Iowa. Elsie Wilson will have his record on a card. W. Albert Noyes, Jr., was a freshman at the University of Illinois then transferred to Cornell College in Iowa, I believe, for the rest of his undergraduate work. So he was here for only one year as a student.
>
> The University of Illinois has had such an unusual record during the last 10 years that this is a good time to comment on it because it is unlikely to be repeated for a long time.

Adams's feeling about the true rewards of his career as a teacher are shown in his introduction of W. M. Stanley for the Willard Gibbs Medal of the Chicago Section of the American Chemical Society in 1947:[28]

> There is no greater personal satisfaction for one in the academic profession than to watch his students initiate and develop their careers. It is, therefore, a special pleasure for me to speak this evening about Dr. Stanley, a student who has demonstrated outstanding ability and has received such well-deserved world-wide recognition for his accomplishments. This pleasure is a very unusual one because Mrs. Stanley, the former Marian Jay, was also a graduate student of mine. Her scientific interest has always meant sympathetic understanding in her husband's work. I am claiming in addition a vested interest in any potential scientists among their four children.

Adams's ties with the university remained warm and close. He gave his personal chemical library, which was extensive and valuable, to the university,[29] and in 1968, to his surprise, the University and the School of Chemical Sciences honored him in a way most fitting to his life-long teaching career by establishing a scholarship fund.

"The Roger Adams Fund was pulled out of the hat a couple of weeks before March 1. I didn't hear about it until then," he wrote.[30] Although he thought it lacked good organization, Adams was enthusiastic about the object: to give a few scholarships each year to promising high school students from the state to attend the university and so attract top younger men into chemistry. Whatever its initial organizational weakness, the fund has continued to attract wide support, and Lucile Adams Brink has acknowledged all contributions in personal notes.

Adams's interests did not include running for public office, but he maintained a good citizen's watch on the political scene. He was strongly opposed to the New

Deal and generally favored the conservative Republican view. He contributed to the Republican party, for which his old friend General Lucius Clay was chairman of the Republican National Finance Committee in 1967. In 1968 Adams wrote Nelson Rockefeller promising his support for the presidency.[31] The reason for his opposition to Richard Nixon is described in detail in a letter to Nixon headquarters:[32]

> I received a notice a day or two ago from your organization, Nixon for President Committee. Although I am a Republican, Richard Nixon is the one candidate who is being considered by the Republican Party for whom I would not vote. This is in spite of the fact that I would do most anything to see President Johnson defeated.

The letter went on to describe in detail an incident that Adams witnessed while he was flying from Taiwan to Osaka in the spring of 1964. Nixon was traveling as a private citizen, although he was accompanied by a large entourage, and, much to Adams's disgust, he demanded special treatment at Osaka. Nixon did not have the health certificate required of all persons deplaning, but eventually he was allowed to land without it, after causing much delay and inconvenience to other passengers whose papers were in order. Adams was both indignant and humiliated for his countrymen by the incident.

When Adams first "retired," he was as busy as he had been before. He wrote,[33] "After I retired as head of the department I took on 4 or 5 activities which I had been pressed to do prior to that time. Then I found that I had too much and still do.... Frankly, I haven't any 'retirement plans' as yet. I have not had time to think about the situation." Marvel wrote in 1958,[34] "I can assure you he [Adams] has not slowed down any as far as I can tell. Sometimes he complains he has, but as near as anyone who lives around him can tell he still does the work of three men every day. Maybe he used to do four men's work and in this way he has slipped a little bit."

Later, particularly after Mrs. Adams' death and the decrease of his consulting and other activities, Adams had more leisure time, some of which he employed in lighthearted correspondence with old friends. He enjoyed for a dozen years occasional joking correspondence with the distinguished (and formidable) Leopold Ruzicka of the Federal Technical Institute in Zurich; it is doubtful if any living chemist wrote Ruzicka in as light a vein as Adams:[35]

> I opened the last number of Helvetica Chimica Acta and found that you had finally reached maturity. Congratulations! I will be there myself in another fourteen months.
>
> All I can say is that if the picture that was used is a recent one, you have retained your girlish figure surprisingly well. I assume you are still gardening and probably are still engaged in some sort of chemical work. I, myself, gave up any attempt at research seven or eight years ago but have been busy on various

boards, most of which do not involve chemistry. At least, my activities have been interesting.

I was glad to have a current photograph so that I will know exactly how you look. I must say that you appear to have grown very serious and might well be a cardinal, bishop, or high priest, or even a candidate to be the Pope. What is your trick to keep your face so free from wrinkles? My wrinkles came twenty-five years ago and have just become deeper and deeper.

Ruzicka's reply (in English) was equally sportive.

By paying ten dollars, Adams arranged for one of his wartime associates, Stanley P. Lovell, to receive a D.D. from the Missionaries of the New Truth, which led to an amusing exchange of letters with Lovell.[36]

Hosts of acquaintances always remembered Adams's birthdays and he received congratulations from friends and organizations here and abroad. On his eightieth birthday on January 2, 1969, he was honored at an open house:[37] "My daughter came down from Minneapolis for my birthday on January 2 and staged an open house which between two hundred and two hundred and fifty guests attended. It was like old times. I have done very little entertaining since Lucile, Sr., passed away." The birthday was also the occasion for a moving account of Adams by his colleague Nelson J. Leonard, in the *Journal of the American Chemical Society*, in which most of Adams's 425 research papers had been published.

As the 1960s advanced, Adams's health remained good in general but his eyesight grew worse, and his doctors required him to give up smoking. He took pride in passing his driver's examination in 1970 and never gave up driving. His physical and mental vigor, judging from his travels and other activities, remained at a high level until his eighty-second birthday in 1971. In February he had exploratory surgery at the Mayo Clinic, which revealed an inoperable malignancy. At his insistence the doctors told him their prognosis: he had about six months to live. Adams intended to keep this information secret, but it leaked out and was soon known to many of his friends and associates. For the remainder of the spring he carried on some of his normal activities, although he refused some invitations, saying his health was not good but not discussing his situation any further, at least in writing.

He put his affairs in order, made sure that his property was properly set up in trusts for his daughter and grandchildren so that his will would not have to be probated, talked with an Urbana minister about his wishes for his burial, and continued some of his outside activities, such as attending meetings of the Welch Foundation Scientific Advisory Board and the Battelle Institute. He continued to play a good hand of bridge.

His associate at Battelle, J. A. Wheeler, wrote:[38]

When I knew his end was near, I made a special visit to him at his home in Urbana. My wife still remembers with what happy memories I returned; he

Illinois University Archives

Adams at eighty.

> and his daughter going over one item after another in the living room to decide what its disposition would be, what friend or family member he would give it to; conversation about Battelle and about Du Pont, and what great organizations they were; memories of Zay Jeffries; and a visit of two neighbor ladies whom he greeted affectionately and to whom—and me—his daughter served tea.

One of his last letters, declining an invitation to attend a retirement meeting at Cornell for Alfred T. Blomquist, a Ph.D. with Marvel in 1932, shows his consideration for others and his courageous acceptance of his illness:[39]

> Gordon Hammes was kind enough to invite me to the occasion to honor you at the time of your retirement on May 22. I have written him that in view of my health, it would be impossible for me to come. But I want to assure you personally that if I were not in bad physical shape, I would be there. I have been very fond of you and have admired your work tremendously. The fact that you were able to go from chemistry into the clothing business for a good many years and then return and do such outstanding chemistry work is nothing short of amazing. Jack Johnson and I often spoke about it and he and I admired you greatly in every way.
>
> I am sorry I won't be able to be there. My condition is not a simple one and the complications that I have are numerous. I simply have had to limit my travelling to the most essential activities, and in fact I am afraid I shall have to give up many if not all of these.
>
> My best wishes to you for the future. I hope that I shall see you sometime or other and have a chance for a good visit.

Adams's last trip was early in June to a director's meeting at Battelle in Columbus, Ohio. He returned to Urbana very ill and was hospitalized immediately, where he was able to have only a few visitors and where he died on July 6, 1971.[40] He was buried in Urbana beside his wife with a graveside ceremony; his friends were asked to send contributions to the Roger Adams Scholarship Fund instead of flowers.

Many of his students were notified by letter from H. S. Gutowsky, head of the School of Chemical Sciences at Illinois; others saw the obituary in *Chemical and Engineering News*.[41] To the large group who had done research with him or had taken his classes, and to the thousands who had known him in his varied activities, his death meant not only the passing of a great man but the end of an era.

Literature Cited

1. Documents in RAA, 40, Krebiozen Investigation, 1951–65.
2. Documented in detail in RAA, 50, University of Illinois Retirement System.
3. RAA, 50, several folders on University Patents, Inc.

4. RAA, 6, Subpoena, to testify in Champaign County Circuit Court, dated October 15, 1968.
5. RA's value as a director of the Champaign bank was emphasized to us by Jack Simpson, formerly with the bank, in conversation on May 23, 1977.
6. Materials for the exhibition were collected and arranged by O. H. Dodson, numismatist, then director of the Classical and European Culture Museum of the university; Dodson, rear admiral U.S.N. (retired), gave us helpful information about the material, and John Edwards of the Champaign National Bank kindly showed us the exhibition at the bank. List of items is in RAA, 64.
7. RAA, 33, Delmont Village, 1967–1968.
8. RAA, 50, Urbana Hotel Co., 1954–1955, contains 10 pages of excerpts from the company accounts and a list of stockholders, both in his handwriting.
9. Conversation with Lucile Adams Brink.
10. Private communication from L. A. Brink.
11. RAA, 5, 1949; RAA, 57, S Correspondence, 1947–9, newspaper clipping from local paper.
12. RA to W. W. Atwood, July 9, 1951, written at Bergen, Norway; NAS Archives, Div. of the NRC; International Relations. In this letter Adams says, regarding a request to serve on UNESCO, "This last year has been so hectic and I have neglected my work so at the University, my better judgment and conscience advise me to avoid new commitments."
13. RA, longhand letter to Gilman, February 23, 1964, giving some details and saying, "I miss her terribly."
14. RAA, 34, European Correspondence R; RA to L. Ruzicka, October 13, 1965.
15. A. H. Blatt to DST, October 14, 1977.
16. N. J. Leonard, private communication.
17. RAA, 36, G.; RA to Mrs. L. A. Green, Sr., of Gallatin, Tenn., July 3, 1968; RA had played bridge with Mrs. Green during his round-the-world trip.
18. Trust fund for Emily: RAA, 16, Adams Personal Correspondence. RA to Emily, September 30, 1966; RAA, 16, Personal, says he will visit Ann Adams in Durham in October; he hasn't seen her for forty years. Other letters to and about Emily: RAA, 16.
19. RAA, 50, U. of N.H.; RA to Ann Adams, October 10, 19, 1966.
20. Correspondence about the Vermont "camp" and his Urbana home: RAA, 16, Personal Correspondence, 1960–67, 1970; also Ref. 10.
21. RA to E. H. Volwiler, September 3, 1965.
22. Text from RAA, 10, Honorary Awards Programs, ca. 1920–70.
23. RA to Karl Heumann, May 16, 1957: RAA, 21, Chemical Abstracts Service; RA to Ralph Connor, May 27, 1957: RAA, 46, Rohm and Haas.
24. RAA, 34, S European Correspondence; RA to C. Schöpf, November 29, 1965.
25. Most of these are listed in N. J. Leonard, *J. Am. Chem. Soc.*, *91*, a–d (1969).
26. Interview taped by C. O. Guss, January 20, 1961.
27. RAA, 50, Illinois Bulletin, 1969–71; RA to W. R. Lowstuter, September, 28, 1970.
28. RAA, 22, Willard Gibbs.
29. RAA, 50, University of Illinois Foundation, Library Gift.
30. RAA, 36, L.; RA to W. H. Lycan, March 12, 1969.
31. L. D. Clay to RA, October 6, 1967, and RA to Clay, October 23, 1967; are in RAA, 44, Nixon Letter; RA to N. A. Rockefeller, June 27, 1968, is in the same file; a newspaper clipping said RA headed a state committee for Rockefeller.

32. RAA, 44, Nixon Letter; RA to Nixon for President Committee, October 18, 1967; the letter is written in blunt language.
33. RAA, 12, McGraw-Hill Encyclopedia; RA to S. D. Kirkpatrick, November 30, 1956.
34. C. S. Marvel to E. H. Volwiler, November 12, 1958.
35. RA to L. Ruzicka, November 9, 1967; and January 9, 1969; are in RAA, 34, European Correspondence R; Ruzicka's answers in the same file. The September 1967 issue of the Swiss chemical journal *Helvetica Chimica Acta* was dedicated to Ruzicka to commemorate his eightieth birthday.
36. RAA, 36, L., Lovell Correspondence.
37. RA to Ruzicka, January 9, 1969; Ref. 33.
38. Private communication, J. A. Wheeler to R. M. Joyce, June 19, 1980.
39. RAA, 36, B.; RA to A. T. Blomquist, May 17, 1971; Blomquist, after receiving his Ph.D., had been out of chemistry for several years during the Depression.
40. Private communication from N. J. Leonard and Mrs. W. E. Palmer.
41. *Chem. Eng. News,* July 13, 9 (1971).

Appendix A

Career Achievements of Roger Adams's Ph.D.s, 1918–58

1918

Ernest H. Volwiler Abbott Laboratories, chairman of the board.

1919

Ralph E. Rindfusz American Writing Paper Co., assistant to the president.

1920

Herbert E. French University of Missouri, professor of chemistry.
Ruth E. Merling Eastman Kodak Co., patent attorney.
Sargent O. Powell University of Washington, professor of chemistry.
Lynne H. Ulich Du Pont Co., research chemist.

1922

Otis A. Barnes Colorado College, professor of chemistry.
Joseph Lowe Hall Kansas State University, associate professor of physical chemistry.
Arthur W. Ingersoll Vanderbilt University, chairman of the Division of Natural Sciences and Mathematics.
John R. Johnson Cornell University, Todd Professor of Chemistry.
Wilson D. Langley University of Buffalo School of Medicine, head, Department of Biochemistry.
Charles S. Palmer Northwestern University, assistant professor of chemistry.
Armand J. Quick Marquette University School of Medicine, professor of biochemistry and head of department.
William Courtney Wilson Pyroxylin Products, Inc., president.

1923

William W. Bauer Pittsburgh Plate Glass Co., chemist.

Waldo B. Burnette United Carbon Co., technical director.
John H. Gardner U.S. Army Chemical, Biological and Radiological Agency.
George D. Graves Du Pont Co., general director of research divisions, Textile Fibers Department.
Russell L. Jenkins Monsanto Co., associate director of Research and Engineering Division.
Ralph J. Kaufmann University of Tulsa, professor and head, Department of Chemistry.
Wilford E. Kaufmann Carroll College, professor, head of Department of Chemistry, and vice-president.
Samuel M. McElvain University of Wisconsin, professor of chemistry.
Katherine Ogden the Liggett School (Detroit).
Charles H. Peet Rohm & Haas Co., chemist.

1924
Herbert O. Calvery Food and Drug Administration, chief, Division of Pharmacology.
Wallace H. Carothers Du Pont Co., group leader.
Emil E. Dreger Colgate Palmolive Peet Co., vice-president for research.
Hermann C. N. Heckel Champion Paper & Fiber Co.
Irvin A. Koten North Central College, professor and head, Department of Chemistry.
John S. Pierce University of Richmond, chairman, Department of Chemistry.
Clifford F. Rassweiler Johns Manville Co., vice-chairman of board.
Nao Uyei Geo. A. Wyeth Research Laboratories, director of research.

1925
Wallace R. Brode National Bureau of Standards, associate director.
Courtland L. Butler, Jr. Army Chemical Center, assistant to director of research.
Ralph A. Jacobson Du Pont Co., research chemist.
John W. Kern Union College, assistant professor of chemistry.
Charles W. Rodewald Washington University (St. Louis), associate professor of chemistry.
Ralph L. Shriner University of Iowa, professor and head, Department of Chemistry.
Florence D. Stouder Scott & Bowne Co., director of research.

1926
Glen S. Hiers Collins & Aikman Co., chief chemist.
Carl R. Noller Stanford University, professor of chemistry.
Shripati V. Puntambekar Department of Chemical Technology, University of Bombay, India.
Jacob Sacks University of Arkansas, professor of biochemistry.

Cyprian G. Tomecko St. Procopius College, professor and head, Department of Chemistry.

William F. Tuley Naugatuck–Rumianca, Italy, assistant general manager.

1927

Merlin M. Brubaker Du Pont Co., assistant director, Central Research Department.

1928

Talbert W. Abbott Southern Illinois University, dean, College of Liberal Arts & Sciences.

James A. Arvin Sherwin Williams Co., director, Resin Research.

Letha A. Behr (née Davies) Columbia College of Physicians & Surgeons, research fellow.

Gerald H. Coleman Dow Chemical Co., assistant director, Britton Research Laboratory.

James F. Hyde Dow Corning Corp.

Lawrence F. Martin Southern Regional Research Laboratory.

1929

Euclid W. Bousquet Du Pont Co., research chemist.

Stanley G. Ford Du Pont Co., director of manufacturing, Organic Chemicals Department.

William H. Lycan Johnson & Johnson, vice-chairman, J & J International, and member of Executive Committee.

Wendell W. Moyer Crown Zellerbach, director of research.

Wendell M. Stanley University of California (Berkeley), professor of biochemistry, chairman of department, and director, Virus Laboratory.

Gail Robert Yohe Illinois State Geological Survey, head of Division of Coal Chemistry.

1930

Martin E. Cupery Du Pont Co., research chemist.

Robert W. Maxwell Du Pont Co., member of Planning Division, Textile Fibers Department.

Horace Avery Stearns Dow Chemical Co., patent attorney.

1931

Louis H. Bock Rayonier of Canada, Ltd., research manager.

Eugene Browning Rayonier, Inc., assistant manager of Technical Services.

Earl H. Johnson Stevens & Thompson Paper Co., technical director.

Leslie J. Roll Eastman Kodak Co., superintendent of Synthetic Chemical Division.

Julius White Montgomery College, professor of biochemistry.

1932

Charles B. Becker PPG Co., Chemical Division.
Edward M. McMahon Tennessee Eastman Co.
Paul R. Shildneck A. E. Staley Mfg. Co., assistant technical director.
Roger W. Stoughton Mallinckrodt Co., research chemist.
Eugene H. Woodruff Upjohn Co., head, Department of Patents & Technical Information.
Han Ching Yuan National Peking University, China.

1933

Chin Chang Fu Jen University, Peking, China.
Joseph B. Hale Eastman Kodak Co.
Ervin C. Kleiderer Eli Lilly Co., executive director of development.
Albert E. Knauf Abbott Laboratories, patent attorney.
Marion M. Davis (née Maclean) U.S. Bureau of Standards.

1934

Shih L. Chien National Taiwan University, Taiwan.
Harold M. Ginsberg had accepted position with Chemical Co. in Buffalo; killed in gas well explosion on honeymoon trip in West Virginia two weeks before job was to start.
Donald F. Holmes Du Pont Co., director of marketing, Textile Fibers, Du Pont U.K., Ltd.
Ching Chen Li National University of Nanking, China.
Wilbur I. Patterson USDA Eastern Utilization Research and Development Division, Agricultural Research Services, director.
Norman E. Searle Du Pont Co., research chemist.

1935

Quentin R. Bartz Parke Davis, laboratory director, Biochemical Research.
William E. Hanford Olin Corp., vice-president and director of research.
Ernest Byron Riegel G. D. Searle Co., director of research and development.
Arthur M. Van Arendonk Eli Lilly, patent counsel.

1936

Bernard S. Friedman Sinclair Research Laboratories, associate director, Organic Research Division.
Chi-Yi Hsing Peking National University, China.
Harold G. Kolloff Upjohn Co., director of supportive research.
Marlin T. Leffler Abbott Laboratories, director of Research Liaison.
Rupert C. Morris Shell Development Co., department manager, Chemical Research and Applications.
Meredith P. Sparks chemical patent attorney.

1937

Elbert E. Gruber General Tire & Rubber Co., vice-president and director of Research Division.

Frank C. McGrew Du Pont Co., technical director, International Department.

Richard F. Miller Eastman Kodak Co., director, Film Technical Services Division.

1938

Darrell J. Butterbaugh Rohm & Haas, president, Micromedic Systems.

George E. Eilerman PPG Co., Fiber Glass Division.

Glenn C. Finger Illinois State Geological Survey, principal chemist and head of Chemical Group.

Allene R. Jeanes USDA, Northern Regional Research Laboratory.

Alva J. Johanson Weber State College, head, Department of Chemistry.

Robert M. Joyce Du Pont Co., director of research, Pharmaceuticals Division.

Edward C. Kirkpatrick Du Pont Co., technical director, International Department.

1939

Theodore L. Cairns Du Pont Co., director, Central Research and Development Department.

William R. Dial PPG Co., assistant director of research, Chemical Division.

1940

Bernard R. Baker University of California (Santa Barbara), professor of medicinal chemistry.

Joe H. Clark American Cyanamid Co., director of Administrative Services, Lederle Laboratories.

Lester J. Dankert Dow Chemical Co., patent agent.

Marvin H. Gold Aerojet Solid Propulsion Co., manager of Propellant Chemicals Division.

Nathan Kornblum Purdue University, professor of chemistry.

Robert S. Long American Cyanamid Co., manager of Commercial Development, Organic Chemicals Division.

Matthew W. Miller 3M Co., group technical director, Photographic Products Division

Frank James Sprules Nopco Chemical Co., technical director, Industrial Research Laboratory.

Howard M. Teeter USDA Northern Regional Research Laboratory, Assistant to deputy administrator, North Central Region.

1941

Arthur W. Anderson Du Pont Co., manager, Scouting Research.

Laurence O. Binder National Science Foundation.

Donald E. Burney Amoco Chemical Corporation, executive director, Amoco Foundation.
John T. Fitzpatrick Union Carbide Chemical Division.
William D. Fraser University of Indiana, professor of microbiology and chairman of the department.
Robert O. Sauer UOP Co., corporate manager, Europe.
Hugh W. Steward U.S. Rubber Co., Naugatuck Division.
Richard B. Wearn Colgate Palmolive International, vice-president of research and development.
Lynwood N. Whitehill Sherwin Williams Co., assistant director, Resin Research.

1942
Alfred A. Albert Hercules Co., director of development, Explosives Division.
Richard F. Phillips Merck Co., assistant manager of contributions.
Carl Mayn Smith 3M Co.

1943
Richard G. Chase General Tire & Rubber Co., technical assistant to corporate director of research.
John D. Garber Corn Products Co., International, corporate manager, Industrial Research and Development.
Robert I. Meltzer Warner Lambert Research Institute, director of chemical research.
Clement W. Theobald Du Pont Co., executive director, Committee on Educational Aid.
Robert S. Voris Hercules Co., manager of Research Systems Group Division.
Joseph M. Wilkinson GAF Co., director of development.

1944
Ming-Chien Chang National Peking University, China.
Charles F. Jelinek FDA Bureau of Foods, deputy associate director of technology.
Robert D. Lipscomb Du Pont Co., research chemist.
Ralph S. Ludington Hooker Chemical Co.
John W. Mecorney Shell Development Co.
Norman K. Sundholm Uniroyal, Chemical Division.
Zeno W. Wicks, Jr. Interchemical Corp., vice-president and director.

1945
Kuang-Hsu Chen National Academy of Peiping, China.

1947
Benjamin F. Aycock, Jr. Burlington Industries, Research Center.
Scott MacKenzie, Jr. University of Rhode Island, professor of chemistry.

1948

Morton Harfenist Wellcome Research Laboratories, acting head, Organic Chemical Department.

Jack S. Hine Ohio State University, professor of chemistry.

Anthony W. Schrecker National Cancer Institute.

1949

Theodore E. Bockstahler Rohm & Haas Co., head, Development Laboratory.

John B. Campbell Du Pont Co., product manager.

Joseph Gordon Lubrizol Corp., director, Research and Development.

James L. Johnson Upjohn Co.

Viron V. Jones American Photocopy & Equipment Co.

Nils K. Nelson Purdue University, Calumet campus, associate professor of chemistry.

Morton Rothstein SUNY Buffalo, professor of biochemistry.

1950

Arnold H. Anderegg Shell Chemical Co., section leader, Synthetic Rubber Technical Center.

John L. Anderson Amtek/Technical Products, director of research.

Bruce E. Englund Kinetics Corp.

Harry F. Kauffman Spoon River College.

Anant S. Nagarkatti Hindustan Steel, Ltd., Calcutta, India.

1951

Edward F. Elslager Parke Davis Co., section director, Organic Chemistry.

Frank B. Hauserman Du Pont Co., director, Technical Services Laboratory.

Karl F. Heumann FASEB, director of Office of Publications.

Raymond F. Mattson SCM Corp., Glidden Durkee Division, manager, Group Market Research.

Kenneth A. Showalter U.S. Steel Co.

Charles N. Winnick Halcon International, assistant vice-president and director of research.

1952

Donald S. Acker Du Pont Co., technical manager, Textile Fibers Department.

Bernard H. Braun USDA Agricultural Research Services.

John M. Stewart University of Colorado Denver Medical School, professor of biochemistry.

Thomas E. Young Lehigh University, professor of chemistry.

1953

Dale C. Blomstrom Du Pont Co., research chemist.

Rolland W. P. Short A. E. Staley Mfg. Co.

Karl V. Y. Sundstrom Svenska Esso A.B., Stenungsund, Sweden.

1954

Richard S. Colgrove Du Pont Co., research chemist.

John W. Way Du Pont Co., research supervisor.

1955

William P. Samuels Union Carbide Co., manager, Carbide Government Research and Development.

Leroy Whitaker Jefferson Chemical Co., patent attorney.

1956

E. L. DeYoung Chicago City College, Loop branch, professor and chairman of Department of Physical Sciences.

Joseph E. Dunbar Dow Chemical Co.

Hugh H. Gibbs Du Pont Co., research chemist.

M. J. Gortakowski Utah State Health Department chief, Chemical Section.

G. R. Johnston Pennsylvania State University, Beaver campus, assistant professor of chemistry.

M. D. Nair CIBA Research Center, Bombay, India.

H. J. Neumiller Knox College, associate professor of chemistry.

1957

James S. Dix Phillips Fibers Corp., project manager.

L. M. Werbel Parke Davis.

1958

W. E. Cupery Du Pont Co., research chemist.

Appendix B

Career Achievements of Roger Adams's Postdoctorates 1936–59

1936

Kenneth N. Campbell Mead Johnson Co., director of medicinal chemistry.
Corliss R. Kinney Pennsylvania State College, professor of fuel technology.
Sidney H. Babcock, Jr. American Cyanamid, manager of Production and Engineering, International Division.

1937

Charles C. Price University of Pennsylvania, professor and head, Department of Chemistry.
Harold R. Snyder University of Illinois, professor of chemistry, associate dean of the graduate college.

1938

Dean Stanley Tarbell Vanderbilt University, distinguished professor of chemistry.

1939

Theodore A. Geissman University of California at Los Angeles, professor of chemistry.
Madison Hunt Du Pont Co., director of research, Pigments Department.

1940

Donald C. Pease Du Pont Co., research chemist.
Edward F. Rogers Merck Co., assistant director of the Research Laboratories.

1941

Cornelius Kennedy Cain McNeil Laboratories, director of chemical research.

Marvin Carmack University of Indiana, professor of chemistry.
Warren D. McPhee Mason, Kolehmainen, Rathburn & Weiss, partner.

1942
Robert B. Carlin Carnegie Mellon University, professor and head, Department of Chemistry; associate dean, College of Engineering and Science.
Kenneth Eldred Hamlin Abbott Laboratories, senior vice-president of Cutter Laboratories.

1943
John E. Mahan Phillips Petroleum Co., director of chemical research.

1944
Stanley J. Cristol University of Colorado, professor of chemistry.
Charles I. Jarowski Pfizer, director of pharmaceutical research and development.
Nelson J. Leonard University of Illinois, professor of chemistry.

1945
Clara Louise Deasy College of Mount St. Joseph, associate professor of chemistry.

1947
Jean B. Mathieu Roussel Uclaf, Paris, France, director of Outside Research.
Allen E. Senear Boeing Co., research engineer.

1948
William Howry Jones Merck Co., research associate.
George R. Thomas U.S. Army, chief, Organic Materials Laboratory, Army Material and Mechanical Center.

1949
T. R. Govindachari CIBA, Bombay, India.
Robert A. Hardy, Jr. Lederle Laboratories, senior scientist.
Werner Herz Florida State University, professor of chemistry.
Wilson M. Whaley, Jr. North Carolina State University, professor and head, Department of Textile Chemistry.

1950
James H. Looker University of Nebraska, professor of chemistry.
Robert M. Ross Rohm & Haas, assistant director of research.
C. R. Walter George Mason University, professor and head, Department of Chemistry.

1951

Eugene J. Agnello Hofstra University, professor and chairman, Department of Chemistry

Joseph D. Edwards, Jr. University of Southwestern Louisiana, professor and head, Department of Chemistry.

Richard Remsen Holmes Hofstra University, professor of chemistry.

Mihoril Prostenik University of Zagreb, Yugoslavia.

Ronald A. Wankel Tennessee Eastman Co., general superintendent, Acid Division.

1952

William Moje Citrus Experiment Station, California, associate chemist.

Irwin J. Pachter Bristol Laboratories, vice-president for research.

Paul R. Shafer Dartmouth College, professor of chemistry.

1953

Seymour H. Pomerantz University of Maryland School of Medicine, professor of biochemistry.

Blaine O. Schoepfle Hooker Chemical Corp., manager of Development and Planning.

B. L. Van Duuren New York University Medical Center, professor of environmental medicine.

Paul L. Vercier Centre Recherche, Cie. Francaise Raffinage, France.

1954

Takeshi Hashizumo University of Kyoto, Japan.

Charles Nelson Robinson Memphis State University, professor of chemistry.

1956

Maurizio Gianturco Coca-Cola Co., assistant to executive vice-president.

Kay R. Brower New Mexico Institute of Mining and Technology, professor and head, Department of Chemistry.

1957

Seiji Miyano Fukuoka University, Japan.

Walter Reifschneider Dow Chemical Co., associate scientist.

1958

H. A. Stingl Toms River Chemical Co., research associate.

1959

Aldo Ferretti U.S. Department of Agriculture, research chemist.

Dragutin Flès Research and Development Institute, INA, Zagreb, Yugoslavia.

Index

A

B

C

D

E

F

G

H

I

J

K

L

M

N

O

P

Q

R

S

T

U

V

W

Y

Production by Janet D. Shoff and Cynthia Hale.
Jacket design by Kathleen Schaner.
Typeset by Circle Graphics, Washington, D.C.
Printed and bound by Maple Press Co., York, PA.